Universal languages and scientific taxonomy
in the seventeenth century

Universal languages and scientific taxonomy in the seventeenth century

M. M. SLAUGHTER

CAMBRIDGE UNIVERSITY PRESS
CAMBRIDGE
LONDON NEW YORK NEW ROCHELLE
MELBOURNE SYDNEY

CAMBRIDGE UNIVERSITY PRESS
Cambridge, New York, Melbourne, Madrid, Cape Town, Singapore,
São Paulo, Delhi, Dubai, Tokyo

Cambridge University Press
The Edinburgh Building, Cambridge CB2 8RU, UK

Published in the United States of America by Cambridge University Press, New York

www.cambridge.org
Information on this title: www.cambridge.org/9780521135443

First published 1982
This digitally printed version 2010

A catalogue record for this publication is available from the British Library

Library of Congress Catalogue Card Number: 81–20147

ISBN 978-0-521-24477-0 Hardback
ISBN 978-0-521-13544-3 Paperback

Contents

Part III The end of the taxonomic episteme

Preface

In the study that follows I describe the whys and wherefores of the rise and fall of the seventeenth-century universal language movement. My thesis is that universal languages were attempts to create taxonomic nomenclatures. They were products of what Michel Foucault has called a taxonomic episteme, which, I argue, is an episteme predicted on a philosophy of Aristotelian essentialism. Taxonomic universal languages came about in response to linguistic inadequacy precipitated by socio-cultural events: the information explosion of the Renaissance and early seventeenth century; the specialization of science; and the development of a literate culture.

I do not claim to give a full history or description of the universal languages. These have been well provided in James Knowlson's *Universal Language Schemes in England and France 1600–1800*. Knowlson's book also provides a compact summary of the scholarly work which preceded it. Nor do I cover in detail all of the various sources in the intellectual tradition; readers desiring a broader treatment are referred to Vivian Salmon's publications, in particular to her introduction to *The Works of Francis Lodwick*. Arthur Padley's more specialized study, *Grammatical Theory in Western Europe 1500–1700*, treats the grammatical tradition to which the universal languages are related and provides a valuable but different perspective – one which is of particular interest to linguists. Insofar as Padley sees the grammatical tradition as Aristotelian, his study is complementary to mine.

The reason I have focused on the scientific background is that for the most part the motivation of the language projectors was more scientific than linguistic; their concern was more with nature than with language. Furthermore, in the seventeenth century, scientific and linguistic theory both derived from an Aristotelian world view in which classification plays an essential part. It seems to me that the development of (biological) science in the seventeenth century lays bare these taxonomic underpinnings and directly reveals their source in Aristotelian philosophy. The nomenclatural needs of the seventeenth century can be understood as a development of taxonomic thinking and of language, and one sees this most clearly through the lens of biological science.

Finally, I have approached the universal languages through the tradition of (biological) science because science provides a view of language which implies that language reflects properties of a real world rather than of the mind. Biology is one of the parts of that real world which serves as a constant reminder that there are inalienable facts of nature, facts that cannot be reduced to merely a mental reality. However much we know and understand the hypothetical or 'irreal' nature of many things we attribute to nature, nothing eradicates or changes the basic fact that natural kinds exist, persist and proliferate. Whether in nature or in language, these appear to be phenomena which must be treated as special cases. Despite any of its misguided developments or applications, taxonomy exists as the representation of these fundamental facts of life.

The other part of this study takes up the individual language schemes. In treating those I am not concerned to give a complete description of the parts. I am rather exclusively concerned with these schemes as (taxonomic) nomenclatures. The reader will note therefore that I treat primarily root words (simple terms) or taxonomic categories. Where I have found it interesting, I include information about the formation of complex terms by derivation. I also treat the written form of the languages where it illuminates the rest of the discussion. I am little concerned with grammatical matters (parts of speech, tense, etc.) and even less concerned with syntax, deep structures and the like.

Since my intent is to relate the universal languages to social, historical and intellectual contexts it has been necessary to impose further limitations. I have chosen to treat the universal language movement primarily as an English phenomenon. This is justified in the sense that the most significant developments occurred in an English setting. Since the universal languages are related to science and specifically to empirical science, I place them in the context of the Royal Society and its prevailing philosophy of science. Also, I have excluded study of Leibniz' linguistic work on a number of grounds. Coming where and when he did, Leibniz had little connection with the English projectors, their scientific milieu, or their kind of universal languages (nomenclatures).

This book was written in Santa Barbara, Oxford, Berkeley, Khartoum, Melbourne, Cairo, Cambridge, Abu Dhabi and Honolulu over an extended period of time, and the debts I have incurred along the way are many. Some very large debts are intellectual ones. It will be clear that Foucault's ideas on taxonomy as a representational system or language are pervasive, certainly more than the sum total of the individual footnotes. I have also benefited enormously from the discussions of two contemporary linguists, Björn Jernudd and Jiří Neustupný. To them I owe awareness of the theory that languages as well as societies go through developmental stages, and that in the process languages are cultivated, treated and planned.

I should like to thank the University of California at Santa Barbara for a summer research grant which enabled me to study material at the Royal Society. I am grateful to Dr Wadad Said for her expert help with Aristotle and to Drs Julian Boyd, Zelda Boyd, Dan Stempel and Reinhard Friederich for reading the manuscript and for commenting on it. I should very much like to thank Dr Martin Hinds and Mrs Pamela Hinds whose hospitality and help in Cambridge made those flying visits both possible and pleasurable. Closer to home, I thank Tor, who put up with a lot and helped even more, always with his unflappable good spirits. And my husband, Björn, of course.

Honolulu, 1981 M. M. Slaughter

Acknowledgments

I am grateful to the following for their permission to reproduce and to quote various materials: the Huntington Library, San Marino, California, for Robert Hooke's illustration of Wilkins' real character from *A Description of Helioscopes* (1676); Lord Delamere and Sheffield University Library for material from Samuel Hartlib's *Ephemerides*; the Royal Society for material from their Letter Books and Classified Papers; the Keeper of Manuscripts at the British Library for material from Sloane MSS 897, 932 and 4377; the Keeper of Western Manuscripts at the Bodleian Library for material from Ashmole MS 788, Aubry MS 13 and Savile MS 4; the author and *Past and Present* for the graph from Lawrence Stone, 'Literacy and education in England 1640–1900', *Past and Present*, 42 (1969) and Roger Schofield for the revised figures in 'Dimensions of illiteracy 1750–1850', *Explorations in Economic History*, 10 (1973).

Introduction

i. Real characters and philosophical languages

In 'The analytical language of John Wilkins', Jorge Luis Borges calls our attention to one of history's curiosities, a Chinese encyclopedia entitled the 'Celestial Emporium of Benevolent Knowledge'. 'On those remote pages', Borges reports, 'it is written that animals are divided into (a) those that belong to the Emperor, (b) embalmed ones, (c) those that are trained, (d) suckling pigs, (e) mermaids, (f) fabulous ones, (g) stray dogs, (h) those that are included in this classification, (i) those that tremble as if they were mad, (j) innumerable ones, (k) those drawn with a very fine camel's hair brush, (l) others, (m) those that have just broken a flower vase, (n) those that resemble flies from a distance.'[1]

The classification is as mad as it is humorous and the celestial emporium from whence it comes, exotic and remote. Closer to home however, what might be surprising is the fact that any number of illustrious members of the nascent Royal Society, in the bloom of the scientific revolution – and this includes Wilkins, its unofficial founder, as well as Boyle, Hooke, Wallis, Ray, Wren, Evelyn, Newton and Leibniz – discussed and produced, with perfect seriousness, similarly enclyclopedic and taxonomic schemes. Not only that, they went a step further and spent considerable time inventing and using equally odd artificial languages meant to represent, with a degree of accuracy unknown since Babel, those cosmic classifications. These projects are what are known in the title of Wilkins' book as real characters and philosophical languages. They popped up like weeds during the seventeenth century, sometimes in the most unlikely of places, starting with the works of Bacon and the letters of Descartes.

At first the language projectors (from *project*) were concerned with constructing a universal writing system, a real character composed with ideographic signs which they believed directly referred not to words but to things or to notions of things, just as 2 refers to the quantity two and not the word

UP

two, *deux*, *duo*, etc.[2] To say that one will invent a common language of ideographic signs presupposes, however, that there is a common set of notions which the ideographic signs are to represent. While these common notions are obvious in the indisputable case of numbers they become more problematical once we go beyond numbers. The language projectors, since many were also scientists, soon made it their business to set out or to discover precisely what those universal notions were.

At first they did not deal with things and notions directly; rather they sought these simple elements in natural languages like Latin, English, Dutch, Greek, that is to say, in 'their' languages. From these they chose a set of terms (e.g. *father*, *house*, *beauty*, etc.) which they considered to be universal, i.e. basic to human thought and experience. The construction of a universal character, it was understood, rested on the construction of a universal lexicon, a set of names which those signs would represent. When the lexicons of root words were put together with the real characters, they produced what are properly understood as universal languages.

Right from the beginning, one of the considerations interjected into the universal language movement was that, whatever the linguistic signs may be (either written signs or words), they must represent not just any notions, but accurate notions. And it was argued, for example by Bacon, that many words existing in natural language were imprecise; they did not clearly represent the real and true phenomena of nature. As the later language projectors opined, an artificially created language could rectify this anomaly between words and things. To use the words of natural language as a base for that language was, however, dangerous. By so doing one ran the risk of dragging traditionally instituted error into it. Following Descartes, projectors like Ward, Wilkins and Dalgarno realized that, in order to have a satisfactory, accurate language, it was first of all necessary to have a satisfactory, accurate account of nature. The proper linguistic representation of nature depended on the proper description or explanation of nature.

This is what defines a philosophical, and not merely universal, language. The creators of the early universal languages made little distinction between primitive words in natural language and primitive or simple elements of nature. The proponents of philosophical languages, however, distinguished between words and the things and notions which words represented. Ideally, they argued, the relationship between the two should be precisely parallel or isomorphic.

It is at this point, where the order of nature is distinguished from the order of language that the influence of science comes in. In order that nature be properly represented in a system of symbols it is necessary that we discover the simple elements on which all phenomena are based, or from which all phenomena are constructed, or by which all phenomena can be analyzed. The analysis of nature into its simple elements or component parts was understood to be the prerequisite of a philosophical language. Constructing it required no less than providing a model or theory of nature. In short, it became a scientific activity. But it became a particular kind of scientific activity: the model and explanation of nature adopted by the language projectors were taxonomic.

ii. Whys and wherefores

There remains the question why. We can undoubtedly explain much of the interest in the universal languages by virtue of their oddness, but, if we take them merely as curiosities, it is difficult to understand the lifelong interest of, say, Wilkins who was hardly a crank, just as it would be difficult to explain the steady interest evinced by modern scholars (which has, if anything, increased). As linguistic enterprises, we know the schemes to be nonsense. Neither man nor language works in ways that are compatible with an artificial language. We are, therefore, brought to ask why did these artificial languages proliferate when they did, what assumptions did they rest on, what did the projectors hope to accomplish, what happened and what did it mean.

The focal point in the language schemes is, clearly, *order* of some sort. The early universal languages attempt to bring order into the linguistic code; the more developed, philosophical languages tackle the order in and of nature. The tables or the taxonomies of the universal languages become the most crucial aspect of the universal language movement.

Scientific taxonomy in the seventeenth century was a method or paradigm bequeathed by and growing out of the tradition of Aristotelianism. Whether its practitioners were aware of it or not and whether they would admit to it or not – indeed, there is every reason to believe that they would firmly deny the relation – it rested on a philosophical view of nature originally articulated by Aristotle. Taxonomic activity in and of itself pre-supposed crucial philosophical assumptions which were in their origin Aristotelian. These include: (1) the belief that nature (reality) is an orderly construct or system; (2) the belief that that system of nature can be known and described; (3) the belief that natural phenomena are comprised of essences which manifest themselves through the genera and species of things; (4) the belief that the proper expression or representation of things, their kinds and their relations is taxonomy (in pre-scientific times, the Great Chain of Being was a construct that acted as a quasi-taxonomy); (5) the belief that definition, which is based on taxonomy, constitutes a statement of essence and that definition is the foundation upon which the structure of certain or scientific knowledge is built; (6) the belief that words (names for things) are, or should be, isomorphic with things, and that there is a nomenclature isomorphic with nature. In the following pages, we shall take up each of these.

Any discussion of classification and classificatory projects in the seventeenth century needs to be placed in the context of seventeenth-century science. We encounter problems here, however, in that most of us understand 'science', the 'new philosophy', and/or the 'scientific revolution' to involve the physical scientists. Taxonomy appears to be, if not incompatible with, at least tangential to science understood in this sense. The difference between physical sciences and taxonomers can be over-emphasized, however, at the expense of similarities and relationships. We shall argue that, in the seventeenth century, taxonomy occurs as one kind of scientific activity, where the aims and intents of the 'new (mechanistic) scientists' and the taxonomers were not, in the beginning, altogether dissimilar. And, as taxonomers, the language projectors shared many of the values, aims and beliefs of the pre-Newtonian scientific community at large. It was not until those values and perspectives changed that the linguists, taxonomers and physical scientists went their separate ways.

When the seventeenth century is described as the age of the new philosophy, this term usually refers to the atomistic–mechanistic theories held by Copernicus, Galileo, Bacon, Descartes, Huygens, Gassendi, Boyle and Newton, to name a few. It is a philosophy deemed to be 'new' precisely because it is said to supplant the philosophy of the Aristotelian tradition (which, ultimately, it did). But while it is true that these scientific revolutionaries depart from the old model in significant ways, we should not overlook the fact that, in the early days of science, certain traditional lines of thought were shared by all, be they new philosophers or not. One such notion was that of the systematicity and orderliness of creation, inherited from tradition and mandated by Christian belief in a rational and beneficent deity. Along with this belief went the concomitant conviction that the system of nature and its ordered principles not only existed but could be known, and known certainly. This was the belief that made early science possible; that provided it with its reason for being:

> The desire for certain knowledge of nature, which inspired Francis Bacon's work on method, and which in fact since St. Augustine or indeed since Plato had inspired the

whole rationalist tradition of European thought, with its belief that what is certain is true of reality, was the principal notion behind all 17th century science; it is what made the 17th century so conscious of method. Until the end of the 17th century, when this Aristotelian form of predication of attributes as inhering in real and persisting substances began to be criticized in the new empiricism of John Locke (1632–1704), all scientists were inspired by the faith that they were discovering through and behind the particular observed phenomena the intelligible structure of the real world. And it was supremely important to have a method that would facilitate this discovery of real nature behind appearances and that would guarantee the certainty of the result. The same emphasis on method is seen in all science, whether in the numerous 'methods' put forward by botanists in search of a 'natural' as opposed to a merely artificial system of classification, or in the experimental method and mathematical method of chemists and physicists.[3]

As Foucault describes it, this universal search for order results in a search for a science of order (mathesis); it is this idea of order that defines seventeenth-century science.[4]

It was an idea that joined Newton and Boyle together with Wilkins in the Royal Society. It was likewise an idea that motivated not only efforts of scientists like Bacon and Newton, Boyle and Ray, but linguists like Dalgarno and Lodowyck as well. Whether they investigated heat or insects, blood or language, whether they composed equations, poems, tables, or sermons, seventeenth-century men were driven by the conviction that there was order at the end of their rainbow.[5]

The newness of the new philosophy called into question the Aristotelian theory of essence and it was usually on that ground that Aristotelianism was repudiated. The degree of that repudiation and the understanding of its ramifications varied considerably from group to group and person to person. But what is significant is how much essentialism survived until the end of the century despite public protest to the contrary. For the mechanistic philosophers the traces are slight, yet nonetheless there. For all but those, however, the struggle to escape from Aristotelianism was much more difficult and much less successful.

For the non-mechanists, particularly for those who studied living things, the belief in Aristotelian essences persisted through the seventeenth century. When they ask what differentiates a plant from an animal, a tree from a shrub, a cucumber from a marigold, they are asking what is the essence or nature of these things such that it is similar to or different from the essence of those other things. (A linguist like Lodowyck is doing the same thing when he puts *moisture, to besprinkle, to baptize, to wet* into the same semantic paradigm.) The notions of genus and species, which in the seventeenth century belonged to ordinary language as much as to natural history, are predicated on the existence of essences. Any utterance or description of nature that invokes these concepts – and they were innumerable in the seventeenth century – implies that what is under consideration is the nature or essence of the thing at hand. It implies that there are essences.

Method arises from the nature of the phenomena being investigated and it is ultimately method which differentiated the new mechanistic philosophy from natural history, differentiated chemists and physicists from naturalists and natural historians. By virtue of their methods the language projectors find themselves allied with natural historians, particularly biologists. What binds the projectors and naturalists together is their common basis in essentialism and their procedures of collection and taxonomy. To the extent that classification was an important activity in the seventeenth century – and it was – we may say that it is a continuation of a philosophical position of essentialism, being drawn still further to its logical conclusion. The language projectors were just one of many groups who participated in the drawing of that conclusion.

Essentialism and natural history find their proper expression in scientific classification whereas the proper representation or expression of mechanistic philosophy is mathematics. Both are images of order, as are the Great Chain of Being and the Newtonian clockwork universe. The one deals with quantities while the other operates on qualities, but both are attempts to create a well-formed representation of the system and order of nature. Mathematics is one kind of representation or language, developed to suit the atomistic–mechanistic theory of nature. Taxonomy is another kind of language or representational system developed to accommodate a more traditional analysis of nature based on essential natures and kinds. Taken together the two methods reveal the common goal of the seventeenth-century virtuosi: the verification of order. To the extent that they are similar sciences of order it explains why it is that both Descartes and Newton tried their hands at philosophical language schemes. It explains further Bacon's interest in artificial languages and accounts for the fact that, in spite of his empiricism and experimental method, and in spite of his rejection of the more mystical notions of Aristotelian form, his scientific philosophizing continually lapses into, or takes the form of, taxonomy. It explains the interests in the artificial language schemes evinced by mechanists like Hooke and Boyle. It also accounts for seeming anomalies on the part of the language projectors: for Ward who apparently saw similarities between atomism, algebra and taxonomy (although he was confused about their reconciliation) and, to take a more trivial example, it accounts for Wilkins, who early on was a mathematician and who wrote, not long after his interest in universal language began, a book entitled *Mathematicall Magick*.

Wherever we turn in the seventeenth century, we see evidence of a desire, need, will to establish the fundamental order of God's creation and of the 'world politick'. Religion and science could go hand in hand because science seemed to demonstrate so clearly the Creator's stable and providential plan for the well-being of man. Science could also suggest a model for a regulated and orderly church and polity, a model desperately needed after the ravages of the revolution and with the beginning of a market economy.[6] It is because of this compelling need to find or assert order that methodology was a matter of such concern.

At its most basic, method functions as a guarantee of order, if in no other way than psychologically. If we can methodize, follow a strict set of rules, consciously proceed by an institutionalized (impersonal) set of directions, we have some sense of triumphing over, or at least controlling, chaos. In this sense science gives the same satisfaction as plays, dances, games. *Which* method guarantees that order is a question that came later, and then mathematical order was deemed to be the more powerful instrument. In the beginning, classification filled the bill.

It is important to understand what early (pre-Newtonian, pre-Lockean) science was, particularly for the English and for the Royal Society. Here I think it is helpful to note that it was called 'natural history' with the emphasis on history. As a historical discipline, it was oriented to the collection of facts about phenomena as a precondition to the end of making generalizations or understanding the phenomena.[7] This is what is meant by the Baconian or empirical orientation of English science. From this perspective, natural philosophy, or science as a demonstrative or explanatory discipline, is best seen as emerging out of, breaking away from, going beyond, transcending, this original disposition towards natural history.

In the period between the formation of the Royal Society and the publication of Newton's *Principia*, natural historians like Wilkins, even Locke, are the norm. Would-be natural philosophers, those willing to entertain hypotheses, like Boyle and to a lesser extent Hooke, are the anomalies. It was not until the end of the seventeenth century that natural history as the paradigm of science was supplanted.

If the first generation of Royal Society scientists held firmly to Bacon's prescription that scientific procedure starts with observation and collection, the idea of collections is interesting in itself. We normally think of collections of things, e.g. butterflies, stamps, old bottles, coins, manuscripts, and of course these are akin to the collections of plants and animals being made by explorers and naturalists, as well as the collections of the Royal Society which needed, as Hooke put it, a repository. It was also in this fashion that scientists thought of data, 'bits' of information similarly treated like things, as for example Boyle's observations on the behavior of saltpeter or Hooke's comments on the 'little animals' seen under a microscope. These are the same kinds of bits that Bacon placed on his tables of absences and presences.

The next problem is what to do with all the bits and pieces once they have been acquired, and probably the most basic and obvious thing is to arrange or classify, to group particulars, reduce the sheer number of individual variations, to generalize. This is the taxonomic step taken by the natural historians. Classification is, however, only half-generalization, local or limited generalization. What comes next is universal generalization – predictive theories, universal laws, explanations that range over all things.

The first generation of scientists were classifiers for a number of reasons. One is that, as a method, classification is natural, by virtue of language and of the mind. Another is that, in the intellectual tradition of Aristotelian logic and of the Great Chain of Being, classification was a necessity. Furthermore, mathematics or their knowledge of mathematics and of the necessity to measure (rather than to describe) was not sufficiently developed. Also to be considered is their reliance on a quasi-organic view of the cosmos deriving from classical times. Combined with this is the fact that, while an amateur can easily follow the botanizing of Ray, it is far more difficult to follow the physics of Newton; organic studies were more familiar; organisms more easily apprehended. A final reason might be that the Royal Society had a 'biological' bias to begin with; the sheer number of physicians in the early Society leads one to suspect that concrete entities, like living things, were simply more to their taste and talents.[8] Living things and taxonomy simply go together.

Although scientific taxonomy as we now know it did not begin to develop until the seventeenth century, it was first discussed by Aristotle and his followers in connection with both logic and the study of plants and animals. In later history, Aristotle's basic classification of substances and of living beings found its way out of the books and into popular culture. For several centuries prior to the seventeenth century it upheld an explanation and schematization of nature known as the Great Chain of Being. If the logic books and the mythology of the Chain provided the major genera of things, obviously the classification could be extended to the infimae or lowest species. The taxonomical efforts of natural historians in dealing with the animal, vegetable and mineral worlds were such extensions. Ultimately, if all the genera in each of Aristotle's overarching categories were subjected to taxonomic analysis, we would have an encyclopedic classification of all known things. It is here that the taxonomic activities of the language projectors fit in. They were, in a sense, the original natural historians of the seventeenth century. Botany, zoology, geology developed as specializations within their original outlines or approximations of taxonomy.[9]

Taxonomy as a science of order proceeds in two ways. First of all, it analyzes wholes into units: constituents, parts, elements, variables, etc. The aim or outcome, depending on the success of the analysis, is the establishment of relations of identity and difference between things. Secondly, taxonomy brings things together in groups and arranges the groups in a hierarchical system. Wholes are decomposed in order that they can be recomposed or reconstituted in a system which expresses the totality of their relations. That totality is expressed in the taxonomic table – in a system that conjoins (and disjoins) things.[10] For our purposes, two points need to be noted here: (1) the isolation and analysis of the individual things and (2) the connecting of many things in a spatial relationship of hierarchy.

In order to perform the first operation, a thing must be removed from the immediacy of its environment, removed from all the contexture, mystery and complexity of life. This operation is akin to removing a word from the immediacy of speech and placing it under the lexicographical scrutiny of a dictionary. The thing, and included here can be a thing like a word, is 'decontextualized' in order to be held up as an object of observation and analysis.[11] We shall suggest in a later chapter that the analogy between words and things is not coincidental. There is evidence to indicate that decontextualization and taxonomic analysis develop with the onset of written and/or printed language and with increasing literacy, where words become physical, visible objects.

Natural history in the seventeenth century focuses primarily on the study of living things, particularly plants (rather than the more complex organisms of animals or microscopic species) because the simple and visible organization of plants renders them more amenable to taxonomic analysis. To the natural historian of the seventeenth century, plant organisms are regarded as structures, i.e. arrangements or configurations of a number of significant elements or variables; the structure is the visible sign of the essence. The isolation of these variables or the decomposition of the organism into these elements permits organic form to be reconstituted or retranslated into a linear language, into a series of successively ordered elements which constitute a taxonomy. Once this is done, they can be represented in natural language and given names.

The work of the language projectors is parallel. In their attempts to create an artificial language, they must take words in isolation from the living discourse of speech. The words are objectified and analyzed into their component parts, i.e. their simple elements, and each element is given a distinguishing mark; they are then recomposed into the abstract paradigms of the taxonomic tables and then into artificial words.

In Aristotelian philosophy, a definition is a particular type of proposition and is a translation of a taxonomic relation into natural language. It is a statement of essence. Insofar as an essence participates in a common nature yet is also differentiated from it, definition for Aristotle is definition by genus and difference. The end of Aristotelian science is the framing of such definitions; it is by means of definition that one gains or captures or explains the nature of things. The end was no different for the seventeenth-century natural historians including the language projectors; it is because their activities were directed towards the production of definitions that they could believe they were doing science.

The articulation of taxonomic concepts in natural language assumes that the terms of the definitions do in fact properly designate the genera and the differentiating characteristics of things. It assumes that words are isomorphic with things. Furthermore, the requirements of logic and of science are such that terms must be unequivocable. In the seventeenth century, however, it was generally felt that natural language was not isomorphic with nature or reality – that words and things did not properly match – an attitude which is not surprising in the multi-cultural, multi-lingual context of Renaissance humanism. In the natural sciences, like botany, for example, terms or names for many species of plants simply did not exist, or often when they did there was considerable variation of names from region to region, language to language.

Two related historical phenomena account for this inadequacy in the ordinary language of the seventeenth century. The first is the information explosion (including science) experienced in the Renaissance. The second factor, which made this information explosion possible, is the development of 'literacy'. From all available evidence, it appears that in the middle quarters of the seventeenth century, England became a literate culture as opposed to an oral one.[12] This is not to say merely that some people read or that some people wrote. It is to say that written language and its media and all the social, cultural and cognitive change that goes with it, became institutionalized and predominant. One of the things made possible by printing/literacy was the collection and transmission of vast amounts of information; this required first that there be an adequate language with which to represent that information. The given language – whether the 'dead' or print-governed language of Latin or the vernacular language of English – was insufficient to fulfill the communication needs of this new stage of economic, social and cultural development. All language needed to be rearranged, readjusted and developed.

Given this lack of the necessary linguistic isomorphism and representation, it was imperative that the scientists create nomenclatures – compilations of words or terms which nominated or designated things. Here we find the explanation for the creation of the taxonomical nomenclatures of the botanists and zoologists; it is no less the explanation for the creation of the artificial languages. Such nomenclatures were the precondition of doing science in the Aristotelian sense discussed above. It remains only to say that, in the seventeenth century, the functions of taxonomical nomination and of definition were combined. The natural historians and the linguists alike attempted to create a linguistic system in which things were simultaneously designated and defined. This was a language, as Foucault has called it, of 'essential nomination'.[13] Others called it in its time a philosophical language.

PART I

The rise of essentialist taxonomy

CHAPTER 1

The Aristotelian origins

i. Essences, genera and species

The history of scientific taxonomy starts with Aristotle and is predicated on a first philosophy of essentialism. Taxonomic principles were laid out both in Aristotle's works on natural history and in his logic. In the logic, taxonomy or the division of things into genera and species is a way of classifying predicates; it is a refinement, if you will, of the ten basic categories of predicates. In the long tradition of Aristotelian logic, the basic idea was developed in detail. Classification of the various categories, say, of substance, took the form of branching tree diagrams that specified the various genera and species of the category. In Aristotle's natural history, as opposed to his logic, the taxonomic notions of genus and species were developed to handle relations between living things, primarily animals. In these writings, Aristotle suggested, quite tentatively, that one might range all creatures taxonomically into a graduated scale of nature. Subsequent commentators combined the discussions from the logic and from the natural history. The idea of a scale of nature had obvious affinities to the classification of the categories, where animals, for example, are one genus of the category of substance and where a taxonomy of animals simply carries out the analysis in that category in more detail. In subsequent generations taxonomy was carried out in accordance with Aristotelian rules, and this method of analyzing nature carried with it at the same time an implicit theory of nature.

That theory of nature begins with matter, distinguished qualitatively into kinds. This follows from, and is in concert with, the general Greek tendency to regard the natural world as an organism: 'An organism implies differentiated organs; in the spherical world-organism of Greek thought there was earth in the middle, then water, then air, then fire, and lastly for Aristotle, the *quinta essentia* of the world's outermost envelope... For Aristotle, nature is made of substances differing in quality and acting heterogeneously; earth naturally moves towards the centre, fire away from the centre, and so forth.'[1]

This qualitative differentiation of the prime matter of the cosmos is combined with Aristotle's notion of hylomorphism, i.e. the notion that 'form' is the active principle of things (substance); its 'spontaneous activity was the intrinsic source of the characteristic and regular behaviour of each natural thing'.[2] Form is what persists through and causes substantial change. 'Matter' on the other hand is the passive principle of things, 'implying the *potentiality* to receive the attributes that become actual with the form'.[3] Ultimately, Aristotle conceived of matter as being 'pure potentiality, capable of determination by any form and having no independent existence. This pure potentiality was called by the medieval scholastics *materia prima* determined by form.'[4] This dual organization of matter, its hylomorphism, is what gives rise to the concept of substantial form or essence. Essence was determined by (the combination of) form manifesting itself through matter.

Aristotelian nature in its most basic form consists of essences of (kinds of) things, which are separated from one another by qualitative differences. The most basic statement of Aristotelian metaphysics concerns what kinds of entities there are and what distinguishes them. At the root of this is the fundamental distinction Aristotle makes between substances and accidents. Substances are those things which exist absolutely (*per se*, independently, in and of themselves). Accidents exist contingently, i.e. their existence requires a prior existent. Thus, for example, man is a substance; it exists independently of any other entity. But white(ness) is an accident; it does not have a separate and independent existence; it occurs contingently with another body. We cannot conceive of the color white without first conceiving of a thing which is white.

While accidents are associated with substances, they are not intrinsic or necessary to them in the way that universals or common natures are. (One needs to be careful here with terminology. We can say that the universal exists in a thing if we understand that this usage of 'in' is different from when we say that accidents are 'present in' subjects in the sense that they are attributes 'possessed by' the subjects.) The quality white is not intrinsic to man in the way that the common nature, animal, would be. Since this characteristic, white, is not necessary to the subject in question it is said to be accidental; it comes and goes depending on the individual and circumstances.

The distinction between substance and accident was set out in the *Categories*, the first book of Aristotelian logic. There is debate about what, ontologically, Aristotle meant these categories to be: a classification of things that exist, or a classification of words or statements that can be made about things.[5] In the seventeenth century, readers understood the categories to be the ten fundamental classes of kinds of entities found in nature. The categories consist of substance (man) and accidents: quantity (two cubits long); quality (white); relation (double); place (in the Lyceum); time (yesterday); posture (sitting); possession (is shod); action (cuts); passion (is cut).

Closely connected with the Aristotelian idea of substance is the notion of essence. Essence is the most general and abstract aspect of entities. It is what permanently makes things what they are and not something else. It is the 'nature' of things. Epistemologically speaking, we know or learn of essence through the substance. Essence is then 'the primary datum of intellect', through which we have knowledge of the existence of things. It is 'that which is placed at the onset before our mind when we form the conception of anything or conceive the idea of it'.[6] But since our intellect apprehends not only (the existence of) the thing but also at the same time the principles or the constituent parts of the thing, there is a stricter definition of essence: it is 'what a thing *necessarily* and *primarily* is as intelligible'.[7]

This concept of essence is still with us in daily language as, for example, when we speak of the 'nature of things', or of 'mankind' or of things being 'essentially' such and such.

Essence is necessarily of a thing. 'It is real necessity (*de re*), which differs both from the ordinary physical necessity of natural laws and from the necessity that is relative to language (*de dicto*).'[8] And because this necessity is *de re*, it is ultimate: once you identify the essence of a thing (as the cause of its essential properties), since essence is indivisible, you simply have no more dividing to do; all things follow from that essence. Essence is the necessary and permanent condition of a thing, what makes it that and no other. Insofar as it is permanent and unchanging it is not applicable to individuals; permanence for Aristotle is only to be found in the common nature or behavior of things.

Since 'essence' and 'substance' both refer to the permanent characteristic of things, they are easily confused. Maritain is helpful here in distinguishing between the two;

> the term substance (*substantia*) corresponds to the Greek word *ousia* understood in a restricted sense. The term *ousia* primarily denotes essence or nature, but *since substances are the first object which the intellect attains* when it considers what exists, they are also for that reason the first object which confronts the intellect with the notion of essence; in other words, they are the first to merit the denomination of essence or nature. Hence the term *ousia*, which taken in its most general sense denotes essences, and is afterwards divided into *substances* and *accident*, has most naturally served to denote in a special sense the first member of the pair, *substance*.[9]

The most important points to note here are that (1) essences are known through substances and are coexistent with substances (no substances, no essences) and (2) essences are intrinsic to a subject or thing, indeed are the necessary and permanent conditions or characteristics of that thing to be that thing.

The category of substance is further divided by Aristotle into primary and secondary substance. Primary substances are the individual entities that exist: Socrates, Fido, this man, that horse, the house. Secondary substances are generic, i.e. they are substances composed of a group of individuals, each sharing a common nature as, for example, man, animal, dog.

Another way of stating this distinction is to say that primary substances (individuals) are not predicated of subjects while secondary substances are. The difference is that 'anything which can be said [or predicated] of something else as its subject must have some kind of generality'. What can be described and articulated about any subject is its less-than-unique characteristics, i.e. more general, commonly held characteristics. The statements we make about things are meaningful only insofar as they refer to common, and not unique, perceptions and experiences. 'An adjective [for example] which could be used only on one unique occasion would not function as an adjective and the something unique it designated would not be sayable of a subject...all things that are predicable of subjects are non-individual. Contrariwise whatever is not predicable of a subject is in some sense non-general, particular.'[10]

Thus there are two features which formally define these notions: 'asserted of', which is a relation of universal to particular and 'present in', which is a relation of attribute to possessor. Primary substance (individuals) is neither 'asserted of' nor 'present in' a subject. Secondary substance (genus and species) is 'asserted of', but not 'present in' a subject. Accidents are 'present in' subjects and may upon occasion be asserted of them (a more general accident may be asserted of a less general one).[11]

The category of secondary substance is the crucial one in Aristotelian metaphysics. What Aristotle claims is that secondary substances are not contingent on (accidental to) prior existents as are accidents; they exist *per se*. And yet, as independently existing things, they are not individual, as are primary substances. Secondary substance is not of one but of many; it is the nature common to many individual things, which is not the same as being an individual or particular thing. A secondary substance like man, for example, does not exist without the individual Platos, Socrateses, etc. And yet man is not individual, not uniquely applicable to Plato and Socrates. The secondary substance is a *kind* of thing.

The question of genus and particularly of species is a complicated one in

Aristotelian logic. In the first place, Aristotle himself used the terms not only in his logical and rhetorical works, but in his biological works. In the second place, the terms were changed somewhat in the ensuing Aristotelian tradition.

The *Topics* is where Aristotle defines his terms. There he lists four kinds of predicables, statements about subjects. A predicate is either a definition, property, genus or accident. (While Aristotle assumes that species and difference are predicables, he does not explicitly discuss or define them.) This list of predicables was later modified by Porphyry to include genus, species, difference, property and accident. I shall use the definitions of Porphyry because they are set out in a more straightforward fashion and because they are what was handed down through tradition.

Porphyry follows Aristotle in defining genus as 'that which is predicated of many things...differing in species in answer to *what* a thing is'.[12] Species is similar. It is 'what is arranged under genus, and of which genus is predicated in reply to what a thing is: moreover, thus species is what is predicated of many things differing in number, in reply to what a thing is'.[13] Difference is what is predicated of many things differing in species in answer to the question, of what kind a thing is. The difference can be thought of as that which divides the genera into species or it can be thought of as that which constitutes the species, e.g. 'rational' divides the genus animal and constitutes the species man.[14] One thing that must be avoided is thinking of the species as the genus plus the addition or attachment of a difference. A difference is a characteristic arising from the essence of a thing; the species is a way or mode in which the genus exists.

For Aristotle, the universals were both instrumental and necessary in reconciling the philosopher's perennial problem of permanence and change. Philosophy, including natural philosophy or science, formulates universal laws, i.e. statements that apply in all cases without exception. This presupposes that at some level nature is regular, permanent, stable and unchanging. If, by contrast, nature were in an everlasting process of change, philosophy and science would not be possible. There would be no regular instances about which to make universal statements. Philosophy and science need, furthermore, to be able to demonstrate the truth of these general statements: their aim is to arrive at certain knowledge. If all is flux, demonstration and therefore certainty is not possible.

For the pre-Socratic philosophers, permanence and change were opposing forces of nature; it was Plato who took the first step towards bringing them together. For Plato nature was divided into two realms: the realm of appearances characterized by change and multiplicity and the world of essential, unchanging Ideas or Forms. Reason told him that these permanent Forms must exist, but it also told him that they exist independently of the nature visible to man. The Forms, he held, exist apart from individual things, exist *ante res*, and things are like images or reflections of the Forms they are associated with. Knowledge of these Forms cannot be attained through the world of appearances; ultimately that knowledge (of the supreme Form) can only be had by means of *noesis* or rational intuition. Because of this it is not capable of demonstration. To Aristotle's way of thinking, however, without demonstration there could be no certainty.

Aristotle's solution was to consider the essences as existing simultaneous to and in connection with, as part of, the visible world. He denied the transcendental realm to which Plato had assigned his Forms, a realm detached, ontologically speaking, from the material world. Aristotle's background as a biologist is not without interest here. Despite the continuous diversity and flux he saw in the world of nature, he saw equally well a certain stability, order, permanence and form.

> Knowledge, Aristotle and Plato agree, is of the finite; only the determinate is intelligible. But Plato seeks finitude, definition against a background of indefinite flux. On the contrary, the finite, the intelligible, Aristotle is convinced, must be sought in an area already bounded. It is all-important for the possibility of science that there are definite things characterized by definite structures and definite life-histories, things which group themselves naturally according to these structures and processes, so that the world, in itself, displays the finitude essential to knowledge, and knowledge mirrors the intelligibility of the world. Not only art but knowledge imitates nature; and what principally characterizes nature is the orderly realization of each kind of thing in its due place and its due time. The unambiguous predications [universals] of science are possible only because things sort themselves out naturally into kinds; knowledge results from the mind's responses to such natural groupings. Transcend them and you transcend the limits of univocal [unambiguous] speech, which are the bounds of science. You have strayed beyond the well-fenced limits of being-what-it-is [essence] and are lost in the quicksands of dialectic [probable, as opposed to certain, knowledge] once more.[15]

Thus it is fundamental to Aristotle's philosophy of science that nature sorted itself into kinds or, more specifically, into genera and species. The existence of these classes in nature is a fundamental principle of Aristotelianism, that which indeed makes philosophy in the sense of certain knowledge possible.

ii. Words and definitions

It was one thing for Aristotle to claim that universals exist *in re*, in conjunction with the phenomenal world. But it was another thing to explain just how they exist, and to prove it. One could, after all, criticize Aristotle, just as one could criticize Plato, on the grounds that these Forms or Ideas or universals were simply bare notions, created by the mind, having no real existence, there being no demonstration to the contrary. But Aristotle (along with Plato) conceived the universals as having a real existence.

It was not long before Aristotelians discovered this problem. In his introduction to the *Categories*, Porphyry raised the question 'about genera and species, as to whether they subsist (in the nature of things) or in mere conceptions only; whether also if subsistent, they are bodies or incorporeal, and whether they are separate from, or in, sensibiles, and subsist about these...'[16] Eventually, these questions were taken up by the Scholastic philosophers in the form of the realist–nominalist debates. The realists were to insist upon (with more or less subtlety) Aristotle's belief in the 'reality' of the universals. Nominalists conceded reality only to individuals (primary substances) and claimed that the universals 'exist' only in and as words – as names designating groups of individuals.

The most general form of the realist position states that there exists a relation, a parallelism, between our notions of nature and nature itself. What we know (certainly) of nature we can be sure is in fact the case and not a metaphor or guess or fantasy. At the heart of this lies the issue of the universals. Realists would claim that (1) we have knowledge of universals, can frame concepts of them and (2) these universals have an extra-mental reality. Aristotle, for example, maintained that the universals exist as a common element of particulars; that a particular is an instance of a universal. For Aristotle, the important thing was that one recognize that universals do not exist prior to things as Plato had maintained.

As for the reality of the universals, Aristotle supposed that they had an objective, albeit abstract, existence: objects belong to real kinds which are there for us to discover and we discover (and name) kinds because they are really there. The ideas we have about the kinds existing in nature correspond to the kinds that are in fact there. Turned the other way around, realists claim that the world of nature, the individuals as well as the universals, can be known. If it exists, it can be known and demonstrated. If it can be known, it exists.

Contrasted with this is the nominalist position. In its extreme form (which it is doubtful was ever seriously held) it claims that there are no universals – either in nature or in thought – except in names or words: 'there is nothing common to a class of particulars called by the same name other than that they *are* called by the same name'.[17]

For Aristotle and subsequent realists, it is not difficult to relate this theory of things to a theory of language. To the extent that words or names of language are made to stand for the things of nature (which are real) and/or our notions of them, language too is real. Thus in Aristotle and among his followers 'words spoken are the symbols or signs of affections or impressions of the soul' just as those affections or impressions are 'representations or likenesses, images, copies' of singular or universal things.[18] Language thus connects with nature and it may be called real.

If, by contrast, there is no relation between nature and our notions of it, and/or between notions and words, then language must be seen as nominal. This is another way of stating the nominalist position that we may express universals in language, but those expressions bear no relation to the nature of things (which is particular). Words then become linguistic conventions, useful perhaps in expressing common experiences, necessary perhaps for communication, but they do not and cannot make claims on reality.

This distinction was maintained rather crudely in Renaissance and seventeenth-century times by the assertion that words are either names of things (that really exist) or names for ideas (which have no counterpart in reality). Hobbes, for example, in the *Leviathan* (1651) took the nominalist line when he claimed that only names are universal, 'for the things named are every one of them individual and singular'.[19]

Some names are *common* to many things, as *a man, a tree*...a common name, being the name of many things severally taken, but not collectively of all together (as man is not the name of all mankind, but of every one, as of Peter, John, and the rest severally) is therefore called an *universal name;* and therefore this word *universal* is never the name of any thing existent in nature, nor of any idea or phantasm formed in the mind, but always the name of some word or name...the conceptions answering them in our mind, are the images and phantasms of several living creatures, or other things.[20]

But seeing names ordered in speech (as is defined) are signs of our conceptions, it is manifest that they are not signs of the things themselves, for that the sound of this word *stone* should be a sign of a stone cannot be understood in any sense but this, that he that hears it collects that he that pronounces it thinks of a stone...nor, indeed, is it at all necessary that every name should be the name of something.[21]

Most of Hobbes' contemporaries took the opposite view. It was indeed so much a commonplace that names represent things or conceptions of things that it is difficult to find a defense of this position. The point is so implicit in Aristotelian logic that defense was considered unnecessary.

In Aristotelian logic, words (names) can be taken nominally or really. Taken alone, they are nominal, that is, they have meaning *qua* words or linguistic expressions, but they have nothing else. 'A noun or verb by itself much resembles a concept or thought which is neither combined nor disjoined. Such is "man" for example, or "white" if pronounced without any addition.'[22] These words express thoughts unaccompanied by truth or falsity so they themselves are neither true nor false: 'a proof of this lies in the fact that "tragelaphos" [literally, "goat-stag", or "unicorn"] while it means something, has no truth nor falsity in it, unless in addition you predicate being or not being of it'.[23] That is to say, words taken alone may have meaning, as 'tragelaphos' has meaning or as 'woman' has meaning (for Renaissance logicians, 'woe unto man') but the meaning is purely linguistic and has no bearing on questions of reality.

Terms used in combination, used with predicates, are another matter. There the use of the term commits one to an implication (or assumption) of the being or non-being of the referent. We cannot say 'Socrates is a man' or 'man is a rational animal' without implying our belief in the existence of a thing coterminous with the term 'man'. How we come to know of that existence is another matter; the point remains that when we use words as symbols for things we assume to exist, we use words really; words in this case stand for (real) things.

According to the tradition of Aristotelian logic, as names can be classed as real and nominal, so too can definitions. The following excerpt from a Renaissance school logic lays bare the distinction with disarming simplicity:

> A definition is two ways considered, for either it is a definition of a word, or of a substance. A definition of a word . . . as, a Realme, is so-called, because it is by a king ruled. . . A woman hath her name so given her, because she bringeth woe unto a man. A definition of the substance, is a speech, which showeth the very nature of the thing, and every perfect definition is made perfect by the general worde and his difference joined together.[24]

Some historians claim that real definitions 'originated' with Plato in the dialogues. The questions raised by Socrates, 'What is justice?', 'What is a Sophist?', etc. purport to be about behavior rather than linguistic phenomena. What is looked for is the essence of the thing, rather than a conventionally acceptable and/or proper use of the term. It is claimed that whenever Plato considers definition he always means real definition, i.e. a definition of what a thing is.

It is to Aristotle that many attribute the distinction between real and nominal definitions. Whether or not he made this distinction, or, if he made this distinction, what he meant by it, is another matter.

Some commentators claim that Aristotle never made the distinction at all:

'everyone, including Aristotle, recognizes that since a definition is a verbal formula a "real" definition is absurd...it leads one into thinking that "real" definitions are only about things and not words'.[25]

It would appear however that Aristotle recognized more than one way of expressing the meaning of names. It is common and no doubt necessary that in the introduction of new and/or technical terms one 'define one's terms'. But definition in this sense is obvious, perhaps even philosophically uninteresting; it is with other kinds of definition that Aristotle is more concerned.

The first kind of definition above we may call nominal, that is, an expression that states 'the meaning of a name, or of an equivalent nominal formula. A definition in this sense tells you, e.g., the meaning of the phrase "triangular character".'[26] This kind of definition runs into the following problems however. If definitions are expressions meaning the same as a name: '(1) both what is not substance and what does not exist at all would be definable, since even non-existents can be signified by a name; (2) all sets of words or sentences would be definitions since any kind of sentence could be given a name; so that we should all be talking in definitions, and even the *Iliad* would be a definition'.[27]

Definition proper must then be something other than the meaning of a name: 'we have a definition not where we have a word and a formula identical in meaning (for in that case all formulae or sets of words would be definitions; for there will be some name for any set of words whatever, so that even the *Iliad* will be a definition), but there is a formula of something primary...'[28] 'A definition is a set of words which is not one by being connected together, like the *Iliad*, but by dealing with one object. – What, then, is it that makes man one; why is he one and not many, e.g. animal and biped...'[29]

The point here is a subtle but necessary one: we can for purposes of discourse define a word by the concatenation of any set of attributes, but by this concatenation we must not be misled into believing we have described a unity, an essence. For to know what makes the unity we must know the cause of the unity. Where there is unity, there is a cause thereof; there is essence and an existing thing. But without the cause there is no statement of essence and hence no statement about an existent. In an obvious way, then, definitions may be of words, but definitions proper are of phenomena, of real things, of say, substances, attributes and events; as such, they are statements of essence.

With definitions proper, we can distinguish between those which are causal, which demonstrate why a thing exists, and definitions of primary premises or immediate terms or first principles, like units in arithmetic. In the latter case, we must assume not only their definition but even their existence and to speak of causes is pointless. The former kind of definition 'explains *why* a thing exists'.[30] This is the same as saying it gives the essential cause of the thing having the attribute(s) that it does have. Thus, for example, a causal definition of thunder would be 'noise due to the extinguishing of fire in the clouds'.[31]

What exactly Aristotle meant to be amenable to causal definition is not altogether clear. Strictly speaking, in a teleological system to ask what any thing is is to ask for its cause. But substances have no causes other than their own forms or 'natures', and it may be that Aristotle had in mind attributes and events which have causes external to themselves.

It would seem clear at least that whatever else Aristotle wished to discuss about definitions, he did have in mind a distinction between (1) the definitions of those things which we know or assume really to exist (hence we can describe the what and the why of them); and (2) definitions about the use and application of terms although these make no commitment as to existence and hence are unable to give any information about the nature of the thing. Simplify this distinction for several centuries and you end up with the distinction between definitions of things and definitions of words: real and nominal definitions, in short, such as we have seen in the Renaissance logic books.

We may sum up Aristotle's philosophy of science as a program for the explanation of nature in three parts: the first part is the essences and their classification and the things which manifest them, i.e. the primary substances which are to be found naturally grouped in genera, the particulars which both participate in the common nature and are individuated from it. The second part is the names of essences, and the third part the statements of essences, or definitions.

Science for Aristotle consists of apprehending essences and constructing statements or definitions of them. Thus definition is central to science, indeed is its end: once we know and can define the nature of the thing, the essence, we have achieved our end. Certain knowledge for the Aristotelians meant certain knowledge of the nature of things, of the essences:

> ...the aim of scientific investigation was to define the substance underlying and causing observed effects...The object of the first, inductive, process in natural science was to define these forms, for such a definition could become the starting-point for the second process, that by which the observed effects were shown by deduction to follow from this definition and so were explained by being demonstrated from a prior

and more general principle which was their cause. The definition of the form was necessary before demonstration could begin because all effects were considered to be attitudes of some substance, and the cause of an effect was shown when the effect could be predicted as an attribute of a defined substance.[32]

iii. Classification in logic and in science

If Aristotelian inquiry begins with the assumption of the existence of essences and ends with the formulation of definitions, what occurs in between these two points is classification. Aristotelian philosophy (and this becomes particularly true in studies of living beings) sees things in terms of qualitative, as opposed to quantitative, differences. It sees things in terms of differences of kinds, natures and essences which are not reducible to the single abstract property of quantity. What seventeenth-century mechanistic philosophy brought was the understanding that the differentiation was quantitative. 'For the new cosmology [beginning with Copernicus] there can be no natural differences in quality; there can only be one substance, qualitatively uniform throughout the world, and its only differences are therefore differences of quantity and geometrical structure.'[33] Phenomena analyzed quantitatively require explanation in mathematical terms. Those divided qualitatively require classification.

In considering these two methods, Whitehead has succinctly stated the difference. The main thing in mathematics, he observes, 'is that in it, we have always got rid of the particular instances, and even of particular sorts of entities'.[34] Consider, for example, the question of number:

> We think of the number 'five' as applying to appropriate groups of entities whatsoever – to five fishes, five children, five apples, five days. Thus in considering the relations of the number 'five' to the number 'three', we are thinking of two groups of things, one with five members and the other with three members. But we are entirely abstracting from any consideration of any particular entities, or even of particular sorts of entities, which go to make up the membership of either of the two groups. We are merely thinking of those relationships between those two groups which are entirely independent of the individual essences of any members of either group...no mathematical truths apply merely to fish, or merely to stones, or merely to colours. So long as you are dealing with pure mathematics, you are in the realm of complete and absolute abstraction...Mathematics is thought moving in the sphere of complete abstraction from any particular instance of what it is talking about.[35]

This is diametrically opposed to the considerations of classification insofar as it is predicated on a recognition and acceptance of essences or kinds. Thus classification, Whitehead concludes, is 'a halfway house between the immediate concreteness of the individual thing and the complete abstraction of mathematical notions'.[36] It is a way of reconciling the general and the more particular aspects of nature: 'the species take account of the specific character, and the genera of the generic character'.[37] Classification results in generalizations of a sort, yet those reach only as far as the kind or essence, i.e. the particular sorts of entities. Insofar as they stop at differences amongst kinds, one can think of taxonomic generalizations as 'incomplete abstractions'.[38]

A table is to taxonomy what an equation is to mathematics, what a proposition is to logic, or what a sentence is to language. Each is an arrangement of isolated variables in a linear sequence or syntax. Each is a set of signs which expresses the relationships constituting the whole. A mathematical or numerical statement is totally abstract insofar as it ranges over all differences in kinds of entities. There are no special cases; number is abstracted from all particular instances. A taxonomic 'statement' is only semi-abstract; it is abstract in that it expresses a general nature which ranges over many individuals; it is particular, however, in that it designates particular, differing kinds of things. Natural language is similarly semi-abstract. Like taxonomy, it makes its divisions on the basis of differences of kinds rather than differences in quantity. If it were not abstract, we would speak with only proper names, names for individual entities and unique qualities. Since a mathematical statement is completely abstract, it has nowhere to go linguistically speaking. The very incompleteness (in abstraction) of the taxonomic concept, however, requires it to be 'translated' into or articulated by natural language.

Incomplete abstractions as opposed to complete and absolute abstractions express the difference between natural language and the language of numbers. To the extent that an analysis of nature is tied to language, and Aristotle's scientific system based on real definitions certainly is, there would seem to be no alternative but to classify.

There are two sources for Aristotle's notions of classification: the first is in the logical works where he lays down the general theory of classification; the other is in the biological writings where he discusses the problems arising in the classification of animals. There is evidence which indicates that Aristotle's early biological studies were instrumental in his developing taxonomic notions as logical concepts.[39] It is also clear in the later biological writings that Aristotle found difficulties in applying the logical notions of classification to living things. It may have been his intention rigorously to apply at some point the logical notions to create a taxonomy of living things but there is also indication that he believed such a classification to be impossible. Thus there is a curious conflict in Aristotelianism between the theory and the actual practice of classification.

In the logical works, as we have pointed out, classification involves the determination of the genera, species and differences of particular things. The genera and fundamental differences, which determine species are predicables, types of expressions which can be predicated of subjects and therefore statements common to many individuals. Genus and species are relative terms, i.e. they indicate a larger and smaller group relative to one another.[40] The highest genus (above which there is no other) is called the summum genus. The lowest species (below which there are only individuals) are called the infimas species. In between are subaltern genera or species. One of these might be considered as a species in relation to the genus above it, or as a genus in relation to the species below it. There was a great deal of confusion in the use of these terms in popular Aristotelian logic.

The process by which genera are distinguished into species is called logical division. In applying this process, a number of conditions must be met. First of all 'it is necessary that the division be by the difference of the differentia'.[41] If we divide animals by virtue of whether they have feet, the next division must be a modification of this characteristic, e.g. whether they are bipeds or quadrupeds, etc. Second, the division must be exhaustive: 'the process wants always to go on so till it reaches the species that contains no differences'.[42] Finally, the end result is the definition of essence: 'If then a differentia of a differentia be taken at each step, one differentia – the last – will be the form and the substance.'[43] We should point out that division is not a discovery procedure. The essence or principles which determine the type must be 'known' *a priori* to the division, in order that the proper difference be identified. The division process selects the proper characteristic which is the expression of these principles. This has led some, like Bacon, to claim that division (and classification) is a process applied to words.

In the tradition of Aristotelian logic, division was the means by which one could frame definitions. It thus took on an important role in providing information about the nature of things, their genera and their species. For the pre-scientific and quasi-scientific world, it was taxonomy.

Aristotle had provided the first division, the summa genera, in his ten categories, and within the category of substance, his cosmology and biology were but further elaborations of division. He did not extend this division to develop a hierarchical arrangement but his commentator Porphyry did and the Porphyrian tree of hierarchically linked genera and species became canonical in the tradition of Aristotelian logic. The tree provides a logical classification of the category of substance as Fig. 1[44] shows:

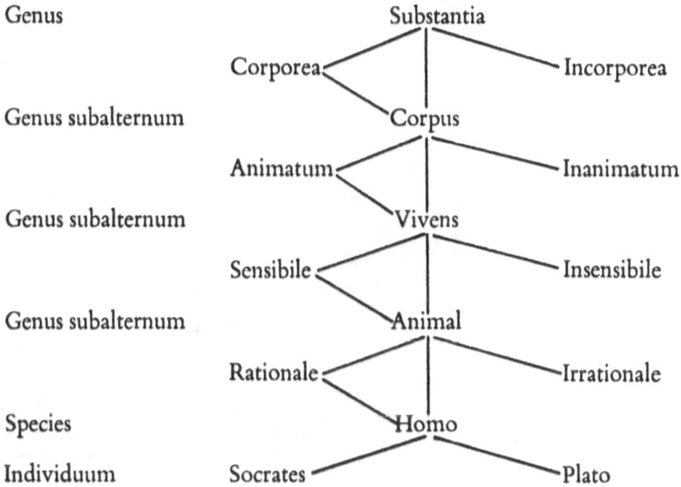

Fig. 1

Porphyry treated only the category of substance but each of the other nine categories can be developed as a genus at the head of a Porphyrian tree. There are then ten possible Porphyrian trees and if each were worked out to its lowest species, we would have an account of all known things. This is what subsequent logicians did. By the sixteenth to seventeenth centuries what we find in Aristotelian logic books is further divisions in each category and in these divisions and tables is included the philosophical and scientific lore of the age. The examples of Blundeville (Fig. 2)[45] and Du Moulin (Fig. 3)[46] are typical of the encyclopedic nature of these logics.

In the theory of classification, it is the criteria according to which the classification is to be made, i.e. what characters are to be chosen as the basis for the division, which are at issue. This issue is crucial in an Aristotelian essentialist framework because the differences must be essential differences; the characters must be the real ones which define the particular essence of the thing at hand.

Crucial here also is the question of the number of characters with which the determination of the species is to be achieved. The question was whether a natural taxonomy could be reached by division by a single character or by a complex of characters. The problem was a serious one for Aristotle and in the early days of scientific taxonomy it lay behind the debates about a natural method versus an artificial system.

The matter is complicated in the works of Aristotle. In his logical works, he argues that classification must be carried out by dichotomous division

- Substance
  - Without bodie as
    - An Angell
    - A spirit or soul separate from the body
  - or With body, Either
    - Simple, either
      - Celestial, as the eleven heavens & all the stars & planets
      - or Elemental
        - Fire
        - Air
        - Water
        - Earth
    - or Compound of bodie & soul either
      - Living, either
        - Sensible Animal
          - Reasonable as Man
          - or Unreasonable
            - A bird
            - A four foot beast
            - A fish
            - A creeping beast, as a worm
            - A snake, a viper
        - or Unsensible as a plant which is either
          - A tree, as
          - A shrub
          - or a herb
      - or Unliving either
        - Perfect
          - Metal, as a stone which is either
            - Gold or silver, etc.
            - Natural, as a Precious stone or flint
            - Artificial, as a tile or brick
          - or Liquor, as wine, honey
        - Unperfect
          - Fiery Impression
            - Thunder
            - Lightening
          - Watery

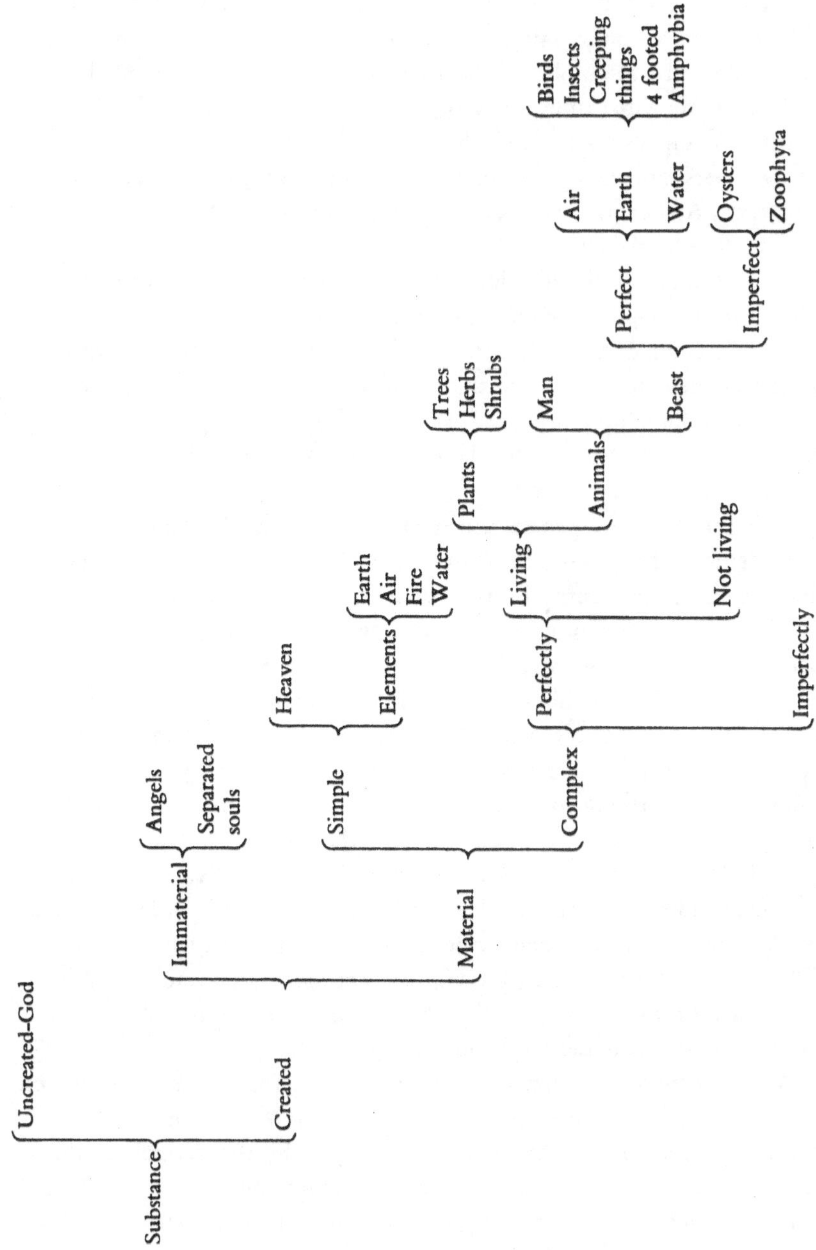

Fig. 3. Pierre Du Moulin, *The Elements of Logick*

until the lowest species are reached. In the biological works, however, where he discusses the actual classification of animals, Aristotle categorically rejects dichotomous division by a single character. Rather in zoological classification, he concludes that a complex of characters (no one of which is essential) must be used. At the same time, Aristotle offered a couple of brief classificatory schemes of animals, each of which proceeded by a single character. And as his successors understood the matter, no doubt from their reliance on the logic texts, Aristotelian classification was to proceed on the basis of division by a single, essential character.

In his zoological classification, Aristotle used primarily morphological criteria. In Aristotelian thinking, the parts or structure of the living thing are the manifestation of the soul, form, or life principle. The structure is dictated by the essential nature of the thing and is the instrument by which the behavior of the thing is produced. From an observational point of view, the behavior is connected to morphology; morphology leads to the genus, species, and ultimately the essence.

In determining the genera and species of animals, Aristotle recognized three degrees of likeness which may obtain between individuals. 'There is first the complete identity of type which exists within a single species.'[47] Here animals are identical in all their parts. 'There is secondly the likeness between species of the same "greater genus": such species have the same bodily parts, differing only in degree – in number, size, softness or hardness, smoothness or roughness, etc.'[48] Here there are corresponding parts which differ in quality; if the parts are different in more than just quality, the animals do not belong to the same genus. 'There is thirdly the likeness by analogy between the "greatest genera" themselves.'[49] This involves a 'morphological correspondence or homology between scale and feather, fish-bone and bone, hand and claw, nail and hoof...'[50] Thus in the animal world, likeness is determined by homology of parts or structure; species or differences are determined by modifications or modes of homology. What is significant for our study is the fact that it is by structure that the nature of things is observed, understood and defined.

When it comes to the actual classification of phenomena, the first division in the Aristotelian cosmology was between the immutable heavens and the changeable, sublunary earth. This was followed by the division of things terrestial into organic and inorganic. Living things were distinguished from non-living things by their ability 'to initiate movement and change without an external mover, that is, [by] the power of self-movement or self-change. The kinds of movement or change common to all living things were growth, the assimilation of diverse matter under the form of the organism, and the continuation of this process in the reproduction of the species.'[51]

Within the category of living things, Aristotle distinguished three groups: plants, animals and men. Plants were defined as possessing only a 'nutritive' soul, i.e. the only activities they displayed were those attributable to all living things – growth, assimilation and reproduction. Animals possess nutritive powers but are distinguished by their additional power of 'sensitivity', i.e. 'the power to respond to environmental stimuli by local motion'. Men possess not only the nutritive and sensitive souls but a rational soul as well, i.e. 'the power of abstract reflection and exercise of the will'.[52]

The sum total of this arrangement was a 'hierarchy of living forms stretching "little by little from things lifeless to animal life", from the first manifestations of life in the lower plants, through the plants to sponges and other sessile animals scarcely distinguishable from plants, and again through invertebrate animals, apes and pygmies to man. Each type was distinct and unchanging, its substantial form being both the efficient and final cause of its particular bodily activity, whether in nutrition, reproduction, locomotion, sensation or reasoning.'[53]

Aristotelian attempts at a more specific taxonomy are not as neat. Aristotle himself dealt only with animals to any significant extent, and rather than producing a single classificatory scheme he suggested several different lines along which a classification might proceed. In the *Parts of Animals* and the *History of Animals*, he suggested a classification which variously uses function, form, internal/external parts, etc. as the basis of division. In the *Generation of Animals* he suggested another classification based on type of reproduction and it is this one which became canonical.

The classification in the *Generation of Animals* chooses as the essential character degrees of 'perfection' as indicated by the degree of development reached by the offspring at birth. This depends on the degree of vital warmth possessed by the parent 'where, through a complex chain of reasoning, warmth is an outward manifestation of the inward essential processes of the animal soul'.[54] Aristotle was impressed by the part played by heat in the hatching of eggs, and concluded it was the agent in all development. The greatest difference in vital heat

> was that which existed between animals with blood (i.e. red blood), and animals with an analogous but colder liquid...The highest types of animal are the vivipara, i.e. those which have enough vital heat to produce offspring qualitatively like the parents ...The next type is that in which a 'perfect' egg is produced, i.e. an egg which does not grow in bulk after being laid. The next large family is that in which an 'imperfect'

egg is laid...Lower creatures have to pass through a third stage anterior to that of the egg and the living young – the stage of the grub...he describes it as turning into a quasi-egg, which is distinguished from a true egg by the fact that no part of it is merely nutriment but the whole of it develops into the living thing. Lower still come the testacea...Some produce asexually a slimy fluid from thich the young develop; in others the young simply bud off from the parents. And, finally, in all the lower types ...there occurs spontaneous generation from lifeless matter such as mud.[55]

		Sanguineous
		1. Man
Viviparous		2. Hairy quadrupeds (land mammals)
		3. Cetacea (sea mammals)
Oviparous	With perfect egg	4. Birds
		5. Scaly quadrupeds and apoda (reptiles and amphibia)
	With imperfect egg	6. Fishes
		Bloodless
		7. Malacia (cephalopods)
		8. Malacostraca (crustacea)
Vermiparous		9. Insects
Produced by generative slime, budding, or spontaneous generation		10. Ostracoderma (molluscs other than cephalopods)
Produced by spontaneous generation		11. Zoophytes[56]

When it came to plants, Aristotle himself discussed only first principles and did not offer any proper classification. The final cause, as we have seen, of all vegetative activity is nutrition and the final end of nutritive activity is reproduction. In contrast to animals, plants are fixed to the ground, hence their food comes to them ready prepared. The nutritive material is absorbed by the roots, travels up the veins of the stem, and eventually produces the fruit or seed.

It was Aristotle's colleague, Theophrastus, who applied these principles to the actual classification of plants. Theophrastus argues that the essential part of plants is the stem, i.e. the organ of nutrition and the material form of the nutritive soul. There are two main groups of plants: the woody, whose stems are long lasting, and the herbaceous, whose stems last for shorter periods, e.g. annually. The woody plants consist of trees and shrubs; herbaceous plants consist of undershrubs and herbs. Each of these is differentiated by the properties (or lack of them) of the stem. Since reproduction is the end stage of nutrition, within each of these four divisions Theophrastus suggests further possible divisions according to the organs of fructification.[57]

Despite the fact that Aristotle came to be known as the father of taxonomy, there was in his zoological writings admission of the impossibility of classification of living things. Although he posits on the one hand a hierarchy of classes which insofar as they are composed of genera and species are clearly delimited and indivisible, he also observed that within nature the distinctions between these classes were not always clear cut. Rather, placed side by side, the classes appeared to differ only by very slight degrees – so slight in fact that exact division was difficult if not impossible to ascertain.

Thus Aristotle observes in the *Historia animalium* that:

> nature proceeds little by little from things lifeless to animal life in such a way that it is impossible to determine the exact line of demarcation, nor on which side thereof an intermediate form should lie...So, in the sea, there are certain objects concerning which one would be at a loss to determine whether they be animal or vegetable. For instance, certain of these objects are fairly rooted [a characteristic of plants], and in several cases perish if detached...Indeed, broadly speaking, the entire genus of testaceans [molluscs, sea urchins, etc.] have a resemblance to vegetables, if they be contrasted with such animals as are capable of progression [locomotion].[58]

> ...nature passes from lifeless objects to animals in such unbroken sequence, interposing between them beings which live and yet are not animals, that scarcely any difference seems to exist between two neighbouring groups owing to their close proximity.[59]

It should be pointed out that Aristotle does not deny the existence of discrete classes in nature. While he maintained that quality was continuous in this formal sense and was therefore infinitely divisible, he never firmly asserted that quality, much less substance, was in the same fashion continuous and divisible.[60] What Aristotle is pointing out is that, depending on the criterion selected for the classification, there will always be some cases in which (1) the objects in question cannot be strictly defined by that criterion or (2) we cannot observe sufficiently the criterion in that specimen. In the first case it is a question of poor definition; in the second case it is a question of poor observation. In either case the groups exist, as must also the demarcations, and classification remains the art of presenting them.

Aristotle's taxonomic work was influential in two distinct ways. In the actual study of and classification of organic things, it provided the method and working principles. As we shall see in the next chapter, when biological taxonomy began to develop in the seventeenth century, it simply picked up where Aristotle and Theophrastus had left off.

The far greater impact of Aristotle's taxonomic notions lay, however, in the use to which they were put for elaborating the idea of the Great Chain of Being. Despite a reluctance to provide a full taxonomy of living creatures,

Lovejoy points out that 'there were in the Aristotelian metaphysics and cosmology certain far less concrete conceptions which could be so applied as to permit an arrangement of all things in a single order of excellence... "all individual things may be graded according to the degree to which they are infected with mere potentiality"'.[61] In the subsequent tradition of Aristotelianism 'this vague notion of an ontological scale was to be combined with the more intelligible conceptions of zoological and psychological hierarchies which Aristotle had suggested; and in this way...the principle of unilinear gradation was added to the assumptions of the fullness and the qualitative continuity of the series of forms of natural existence'.[62]

The result of these notions was the conception of the Great Chain of Being:

> the conception of the plan and structure of the world which, through the Middle Ages and down to the late eighteenth century, many philosophers, most men of science, and, indeed, most educated men, were to accept without question – the conception of the universe as a 'Great Chain of Being', composed of an immense, or...of an infinite, number of links ranging in hierarchical order from the meagerest kinds of existents, which barely escape non-existence, through 'every possible' grade up to the *ens perfectissimum* – or...to the highest possible kind of creature, between which and the Absolute Being the disparity was assumed to be infinite – every one of them differing from that immediately above and that immediately below it by the 'least possible' degree of difference.[63]

The qualitatively determined forms or essences were so arranged to form an ordered, systematic, hierarchically arranged cosmos. We see this notion articulated by no less than the historian of the Royal Society:

> Such is the dependence amongst all the orders of creatures; the inanimate, the sensitive, the rational, the natural, the artificial; that the apprehension of one of them, is a good step towards the understanding of the rest. And this is the highest pitch of humane reason; to follow all the links of this chain, till all their secrets are open to our minds; and their works advanc'd, or imitated by our hands. This is truly to command the world; to rank all the varieties, and degrees of things, so orderly upon one another... that standing on the top of them, we may perfectly behold all that are below...[64]

The Great Chain of Being is not a taxonomy in the scientific sense of the word; rather it can be seen as a highly elaborated folk taxonomy, and as such a predecessor of scientific taxonomy. When the time came in the seventeenth century for scientific investigation to supplant, or at least to supplement, folk taxonomy, the legends and lore of the Great Chain had little to offer which was concrete other than language. More importantly, what the construct of the Great Chain did was to provide a belief in the systematicity, the ordered regularity, the rational design, of nature – a systematic ordering of all the many and various elements of creation. What is particularly important here is the belief that the structure or the system was real and not a bare notion; that the divisions of nature were hers and not ours. This is confirmed by, is indeed rooted in, the structural morphology of living things. Insofar as that orderly system was real and intelligible, it could be known with the certainty required of science; technically speaking, it could be classified, named and defined.

CHAPTER 2

Taxonomy

i. Socio-cultural change, literacy and the development of science

Originally, the Aristotelian world view, retained and popularized in the concept of the Great Chain of Being, encompassed all phenomena of nature – everything embraced hierarchically from the heavenly planets to the lowliest worm and the mud in which it burrowed. Mechanistic philosophy made inroads on the higher reaches of the heavens but for the most part it left untouched the world of living creatures.

Aristotle had posited that animate and inanimate nature are two fundamentally different things. If they are, there are then two separate and distinct laws of nature governing the operations of each, and we cannot apply the theories of inanimate matter to living creatures. It follows that neither can we apply the same methods of analysis.[1] Mechanistic philosophy challenged this assumption by arguing that animate and inanimate matter were in fact composed of heterogeneous particles and hence were governed by only one set of laws. But this did not become fully explicit until the end of the seventeenth century, a subject which we will treat below. Until this fundamental issue of metaphysics was decided, taxonomy would hold a pre-eminent position in the study of living things.

While the most dramatic intellectual and scientific activity of the seventeenth century concentrated on inanimate nature, it should be pointed out that an equal amount of attention was being devoted to living organisms. An indication of the importance of organic studies is Merton's analysis of papers presented to the Royal Society between 1666 and 1702. If we include under the rubric 'organic' not only the biological sciences, botany and zoology, but also anatomy, physiology, pharmacy, pharmacology and medicine, we see that these represent 44% of the total number of papers. The physical sciences of astronomy, physics, chemistry and technology represent 38% of the papers. If we broaden the scope of these inorganic studies to include logic, epistemology and mathematics, the total is raised to 43%. Although the figures pertain to the latter part of the seventeenth century, there is no reason to suppose that the proportions for organic studies are not similar, or even greater, in the earlier part of the century.[2]

It is in the study of living things that Aristotelian essentialism persisted relatively untouched. This means that in the outburst of scientific activity concerned with natural history (and we include here the activity of the language projectors) we should expect to find classification, the appropriate method of natural history, playing a major role.

Broadly speaking, classifying is part of thinking and classification is inherent in natural language; because of this we cannot speak of classifying or not classifying (we *will* classify) but only of different kinds of classification.[3] The kind we shall investigate here, the kind that was developed in the seventeenth century, is logical scientific classification which results from analytical procedures formalized by (Aristotelian) logic.

It has been argued that logic and logical classification, as well as the philosophy on which they are based and the science which proceeds from them, can be associated with literate culture or with the existence of written language and the products and institutions that accompany it; that writing encourages the kind of analysis developed by logic, and provides the means that make logical classification possible.[4] According to this line of thinking, 'it is not accidental that major steps in the development of what we now call "science" followed the introduction of major changes in the channels of communication'.[5] In Sumer it was the introduction of writing, in Greece, the development of the alphabet, and in Western Europe, the invention of printing.

We would expect that cognitive changes brought on by these changes in media not only affect the study of nature but also affect language – attitudes towards language and the use of language. This peculiar conjunction of scientific and linguistic activity is what we find in the case of the artificial language schemes.

This line of thinking suggests further that the primary event or change, brought about by writing and reinforced on a wide scale with printing and the increase in literacy made possible by printing, is 'decontextualization'. This is the separation of language (speech, discourse) from a living context, a context of ongoing activity in which speech is one of many simultaneously occurring events.[6] Writing separates speech from a sequence of time where speech is transitory, and changes language into something static. It separates messages from face-to-face interchanges where personal considerations are negotiated at the same time as information is being delivered. (Modern bureaucracy, with its dependence on the memo, notice, recorded message, files, etc. is the exact opposite of this.) It separates what is said from stimuli in the immediate environment: the stimuli, for example, of questions being asked, comments being made, objections being raised, an interruption, a demand, a request, etc. It separates speech from a scene in which all sensory apparatus is (potentially) engaged, where all the human faculties of the participants are in use. Once removed, it separates the function of listening from the function of remembering. There is no doubt more.

The immediate consequence of this, it is further argued, is a kind of 'fragmentation' whereby the original unity of language – language as part of the totality of nature and events, as participant in and with events – is broken. Language is decontextualized from life.[7]

Writing separates language from a temporal context. It stops the flow of time to which speech is subject and puts it in a permanent form. It transposes language from an aural, temporal dimension into a visual, spatial one. What this transposition permits is a kind of attention to and inspection of the message, of the language and the words, which is not altogether possible in a spoken situation. Once in writing, utterances may be contemplated in part, in whole, in any order; words may be isolated and analyzed. Individual words may be considered in isolation (a situation that is rare in discourse: it is difficult to discuss or to analyze words in speech). Words may themselves become the subject of discourse.

Once we can stand back from our speech acts, once we can isolate language and words and study them as texts, we have the conditions under which language can be analyzed in an abstract, generalized, formalized, rational fashion. It is with this that logic, formal philosophy and science, or the ways of thinking which lead to them, begin.

The reduction of speech to a visual form permits both objectification and abstraction. Once speech is 'translated' into writing, language or words cease to be events and become, instead, things. They are objectified and in this they lose much of their particularity, their intonation, their associative power, their concreteness. Similarly, once this 'objective' habit of mind is fostered, it changes our apprehension of nature. Like language, nature can then be apprehended as a series of separate and isolated items, detached from the 'action' and processes of life, from the living environment. The difference is the difference between the birds in their natural habitat – including the sky, the weather, the trees, the songs they are singing, the berries they are eating, the way they are moving, associated with this time and this place – and the same birds in a museum case. It is the difference between the lilies of the field and potted plants. It is this way of looking at things, this decontextualization and objectification encouraged by literate culture, that fosters logical, rational, scientific activity.

In an interesting series of publications, Havelock has traced the evolution of Greek knowledge from the epic narratives of Homer to the highly abstract philosophizing of Plato and Aristotle. What he argues is that there is a correlation between the style and form and the introduction of (alphabetic) writing together with the changes it brings to ways of thinking and representing fostered by oral culture.[8]

One of the primary differences between oral and written discourse can be seen in their relation to memory. Oral discourse is essentially mnemonic; it is constructed along principles which foster memorization and it is the medium in which the lore of the tribe is preserved. Written discourse relieves the mind of remembering and preserving and assigns these tasks to a repository on paper. Thus, Yates' study of the art of memory reminds us that, while some men in classical and medieval times were literate and while written language and its particular forms of thought flourished in some situations, many were not, and literate culture developed side by side with oral culture.[9]

The medieval scribal tradition offers a curious blend of features from oral and literate cultures. We might think of scribal manuscripts as a half-way house between the literate and the oral. Much of what is presented in medieval manuscripts is folk wisdom – things perceived in the fashion of oral culture (concrete, embedded in time) but organized and recorded more in the fashion of literate culture. It is as though Plato were inscribing the oral world of Homeric man. Furthermore, we know that medieval man quietly spoke the words as he read, making reading a curious form of speech. The proliferation of (medieval) mnemonic treatises only makes sense in terms of an oral culture; the art of memory is obsolete once communication passes exclusively into written form. (Mnemonics were used primarily for pedagogical purposes in the seventeenth century by those who could not, or who were learning, to read. Bacon had taken pains to deny that the mind could organize material deposited in the memory with any degree of accuracy; accuracy for him required written data and tables.)

The medieval scholarly tradition, on the other hand, is clearly not all superstition and fragmentation. This was the age that produced scholastic philosophy and logic, which in turn fostered page-oriented forms such as diagrams and visual models.[10] It was also the age which produced the first encyclopedias. Thus, C. S. Lewis distinguishes the literate medieval scholars as classifiers *par excellence:* 'at his most characteristic, medieval man... was an organiser, a codifier, a builder of systems. He wanted "a place for everything and everything in the right place". Distinction, definition, tabulation were his delight...There was nothing which medieval people liked better, or did better, than sorting out and tidying up.'[11]

In many ways, the concept of the Great Chain of Being is a perfect example of this medieval blending of the oral and the literate. The classification of groups into an all-inclusive cosmological hierarchy is a kind of abstract thinking fostered by literacy and indeed the elaboration of the concept originates with the literate Church Fathers. On the other hand, the image of the chain itself, like the places of the memory systems, is a handy mnemonic device for remembering the outlines of the classification and remembering where to put what. It is also the concrete kind of device suited to oral transmission of information as in, say, a sermon.

When we say, therefore, that decontextualization of language and of nature was taking place in the seventeenth century, we do not mean to imply that this had not happened before. What did happen in the seventeenth century, however, was widespread dissemination of printed matter and, even more important, widespread literacy. With that came widespread decontextualization.

Precise figures on literacy are hard to come by, not in the least because the term itself is so difficult to pin down. Nevertheless, what is clear is that the first three-quarters of the seventeenth century witnessed very active growth in the rates of education and literacy:

> The first two hundred years after the introduction of the printing press saw a rapid growth in literacy, so that by 1675 the national average [in England] for men was probably not far off 40 per cent, ranging from up to two-thirds for the cities to as low as one-tenth in the remote rural backwaters of the highland zone in the north and west. The achievement of such a figure in so short a time postulates a rate of growth over the previous century which is almost comparable with that of the Victorian Era.[12]

By the late seventeenth century, for the first time in history, written ways of knowing and representing were on an equal footing, culturally and socially speaking, with oral discourse (see Fig. 4).

In the domain of language and in the domain of nature, extensive decontextualization was clearly taking place in the seventeenth century. We see it most primitively in the enumerative lists being compiled, in the collections of instances and specimens that underlie, for example, the Baconian methods of induction as well as the collections of plant, animal and mineral specimens that abound. One of Bacon's projects was the institution of Solomon's House, a spatial repository for these collections of decontextualized items. Tübingen

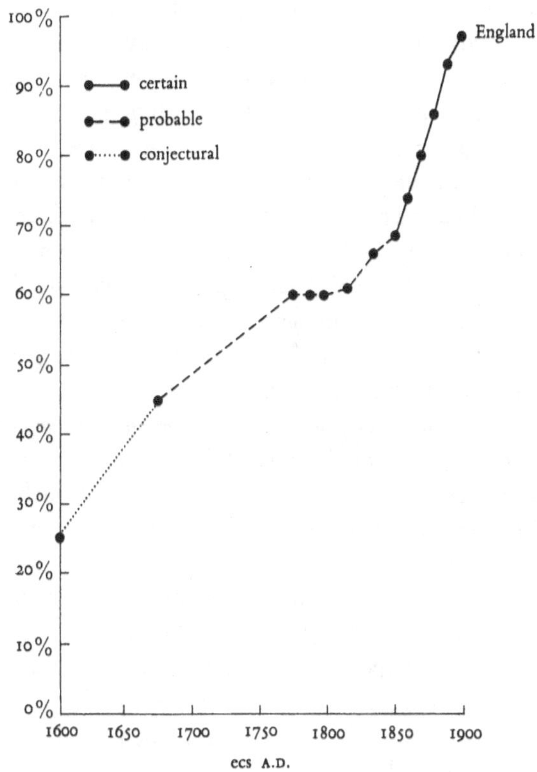

Fig. 4. Estimated adult male literacy in England with Wales 1600–1900[13]

University was described as a 'collection of universal knowledge'.[14] Decontextualized items were put into another kind of repository, into books: in, for example, the many encyclopedias, lexicons and herbals that were being produced. In books themselves there was decontextualization in the increasingly frequent compilations of indexes and bibliographic references.

ii. Decontextualization and written language

A decontextualized nature was treated as a collection of physical things. This can be seen in the textbooks where there is a tendency to reduce the subjects of natural philosophy to single, concrete objects. Universals, for example, are consistently 'objectified'; there is a 'rationalistic tendency to reify the universal natures and to make these reified universals the primary objects of scientific knowledge'.[15]

This tendency towards the isolation and objectification of elements affects not only natural entities but also linguistic entities (considered to be part of nature). One study of seventeenth-century language treatises finds that in these works there is an assumption that 'the singular most important part of any system are the elements'.[16] These elements might be, for example, individual words as opposed to words in a syntactic sequence. The importance of isolated words is manifest not only and obviously in the artificial language schemes but also in teaching grammars which follow the Comenian tradition. In these grammars language study beings with learning and understanding lists of individual words as individual signs for individual things rather than strings of words in sentences or discourses. The elements might also be the individual sounds of which words are constituted. This particular kind of isolation is manifest in the many analyses and classifications of sounds, e.g. Robinson, *The Art of Pronunciation* (1617); Wallis, *De loquela sive sonorum formatione tractatus grammatico-physicus* (1653); Holder, *Elements of Speech* (1669); Lodowyck, 'An Essay Towards an Universal Alphabet' (1685); parts of Wilkins' *Essay* (1668). Another source is the many orthographical treatises of the times.[17]

In the study of grammar, Padley notes the decline of sixteenth-century humanist tradition where the aim was to describe language based on discourse and usage in the speech of the most cultivated (Latin) authors. In its place came a grammatical theory (derived from the Scholastics) which treats language as a thing or substance and as such submits it to Aristotelian analysis based on metaphysical principles. We might also point to the fact that in all grammars the most attention is paid to the (individual) parts of speech rather than to syntax.[18]

Not only does decontextualization objectify but it also abstracts. It transforms unique, living items, proper nouns as it were, into common nouns. The only way to communicate (about) a particular item and at the same time to capture its uniqueness is to point, something that can be done only in the living speech situation. Any form of *representing* the object necessarily abstracts. Some forms abstract more than others – a photograph may be said to abstract less than a detailed drawing and both abstract less than words. The very notions of 'form' and 'essence' are highly abstract modes of thought and represent a highly abstract manner of conceiving of things. It is not coincidence that Plato's *eidos* (idea, form) comes from the same root as 'to see', the visual rather than the oral–aural being the primary mode in which a literate culture functions. The highly abstract metaphysics of Plato and Aristotle, and particularly the Aristotelian kind of essentialism that sees phenomena as essences revealed through structures can be said to be fostered by literate culture and its decontextualization. The continuance of essentialism in the highly literate culture of seventeenth-century western Europe would seem to be quite normal.

It is beyond the scope of this study to show the way in which essentialism as a particular kind of abstraction runs through much of seventeenth-century science. To provide an exhaustive list of instances would be a monumental task. That it can be found even among the unlikely mechanists gives some idea of its extensiveness. As Popper argues, so prevalent was the philosophy of essentialism that it is 'part of the Galilean philosophy of science [and] Newton's laws of motion simply describe in mathematical language the state of affairs due to the inherent properties of matter; they describe the essential nature of matter'.[19]

Another good indication of its persistence is found in school text-books. There we see that 'in its broad outlines the world view of the XVIIth-century text book writers is still fundamentally Aristotelian'.[20] 'The matter-form theory constitutes the fundamental core of text-book natural philosophy. There is some recognition that this theory is under attack, but the objections are dismissed as trivial or absurd. All the manualists without exception hold that the essence of every material body is composed of a passive determinable principle, prime matter, and an active, determining principle, substantial form.'[21] We might add that the attack was delivered to Aristotelian essentialism by various theories of atomism; these were mathematical in nature, and clearly instances of a different kind of abstraction.

If the seventeenth century inherited a tendency to conceive of (natural) entities in an abstract fashion, we can see the particular ways in which these abstract natures and qualities were perceived in contemporary botanical discussions of 'structure' and 'character', the visible manifestations of essence. Although a living organism is a highly complex arrangement and interaction of organs, acting, functioning, and changing through time, the biological concept of structure reduces the organism (and its essence) to a static and abstract arrangement or configuration of the separate parts and their qualities. Structure 'reduces the whole area of the visible to a system of variables [form, size, number, arrangement, etc.] all of whose values can be designated, if not by a quantity, at least by a perfectly clear and always finite description'.[22] In the seventeenth century it was believed that this structure was made visible by means of a 'character'. As a word is a sign of a thing, a character, strictly speaking, is a sign of a structure or a part thereof. It is needless to say that in the process of being rendered as a structure, the organism loses a lot of its uniqueness, not to mention a lot of its 'life'. We will take up the subject of structure in detail when we come to discuss the taxonomy of plants.

Writing and print are communication media which engage us visually as opposed to orally, tactilely, etc. The existence of this visual mode, or the presentation of language in this visual mode, fosters change in our ways of knowing and in turn our ways of representing what we know. With the introduction of written media, we come to apprehend the external world visually. Written language fosters a conception of things primarily in terms of what is seen:

> Observation, from the seventeenth century onward, is a perceptible knowledge furnished with a series of systematically negative conditions. Hearsay is excluded, that goes without saying; but so are taste and smell, because their lack of certainty and their variability render impossible any analysis into distinct elements that could be universally acceptable. The sense of touch is very narrowly limited to the designation of a few fairly evident distinctions (such as smooth and rough); which leaves sight with an almost exclusive privilege, being the sense by which we perceive extent and establish proof, and in consequence, the means to an analysis *partes extra partes* acceptable to everyone: the blind man in the eighteenth century can perfectly well be a geometrician, but he cannot be a naturalist. And, even then, everything that presents itself to our gaze is not utilizable: colours especially can scarcely serve as a foundation for useful comparisons. The area of visibility in which observation is able to assume its powers is thus only what is left after these exclusions: a visibility freed from all other sensory burdens and restricted, moreover, to black and white.[23]

Moving from theory to practice, we see a seventeenth-century school grammar claiming to 'Facilitate the Education of Youth by Bringing Down the Rudiments of Grammar to the sense of Seeing'.[24]

Writing and print also open up new forms of visual and graphic representation. Among the various forms of written representation which we see in the seventeenth century, we can include the simple ones like lists – lists of words or lexica, lists of things, lists of events like astronomical data, or chronicles or genealogies. A different kind of list, a morphological one, would be the paradigm, as in the paradigms of Latin verbs. Recipes are a similar kind of list of ingredients and/or directions. Commercial transactions employ the kind of lists that are the account and the bill.

From here we can go to more complex or developed forms of graphic representation, each varying but each sharing a common foundation in writing/printing. We have more specialized 'lists' like dictionaries, encyclopedias, herbals, etc. We have more abstract, single strings like equations, formulae, propositions or these put together in syllogisms.

New kinds of graphic representation are opened up specifically to language in the seventeenth century. Here we can point not only to a growing concern with orthography, but to the invention of shorthand and to the invention of sign language for the deaf and dumb, as well as to the invention of real characters.[25]

Another, obvious kind of graphic representation, combining the vertical expansion of the list with the horizontal expansion of the formula, is the table. Tables abound in the literature of the seventeenth century, from the tables of topoi and commonplaces to those of mnemonic systems, to those of the logic books, to those of anatomies, isogoges, natural histories, and so on. Many will recall here the innumerable tables of Ramus.[26] The table is the graphic representation of taxonomic discourse, just as written words are the graphic representation of spoken discourse; and just as writing/printing are graphic representations of speech, so the table is the graphic representation of classification and of the taxonomic structure.

Decontextualization of words and of things in nature permits analysis and it also permits arrangement. To some extent the two words 'analysis' and 'classification' are slippery since we tend to think of analysis as taking apart, whereas we think of classification, at least in its end product, as putting things together (in a set of relations). In the Aristotelian, logical tradition, however, analysis and classification are part of the same operation.

What is meant by classification in Aristotelian logic is determining the species of a genus, where genus and species are logical terms referring to *any* superordinate class and its members. The categories of kingdom and phylum in modern biology would be a logical genus and species respectively. Since the species of a genus may, in turn, be regarded as a genus relative to the species' subordinate members, these logical concepts imply a hierarchy of linked categories; they are sets within sets within sets. The method whereby this hierarchy is created is called logical division, i.e. the genus is divided into its species by virtue of a fundamental difference. Since the species is a mode in which the genus exists, to isolate the species one must determine the genus and the fundamental difference. In taxonomic analysis, these are the constituents of a thing.[27]

It is to be noted that Aristotelian classification performs therefore two functions. It analyzes the constituents of a thing and by so doing relates the thing to the next most inclusive category. It combines and arranges these categories of increasing inclusiveness in a hierarchical system in which each category and its members have a specific place. By virtue of the fact that the categories are linked, an individual is related not only to its superordinate group, its 'kind', but to the entire hierarchical system. It is this elaboration of the arrangement into a *complete* system which distinguishes logical classification from folk taxonomy. In folk taxonomy, there is some degree of arrangement, but it is often partial and not worked out in detail, i.e. not worked out from the highest category down to the infima species with no gaps. Furthermore, the fact that a scientific, Aristotelian taxonomy completely spells out relations between individuals and groups, as well as between groups and superordinate groups, i.e. spells out all possible relations, means that this kind of classification provides a complete set of definitions of all possible things. This is what links taxonomy with the emergence of specialized scientific language or terminology and what links it with the artificial languages.

iii. The development of scientific (botanical) taxonomy

When we speak of scientific classification and taxonomy in the seventeenth century we are speaking primarily about the classification of plants and animals, particularly about plants, since they are the simpler organisms (their organs are more visible) and hence lend themselves more readily to analysis. If we trace the development of plant taxonomy as one of the primary examples of natural history, we see that this natural history was the context in which the artificial languages developed. It is no coincidence that many of the figures connected with the artificial languages were also connected with natural science. Natural history provided the methods of constructing taxonomies and the assumptions upon which those methods were built. The search for order was being conducted in all areas of nature and the methods of searching were freely exchanged between one part and another. We shall look at this transmission of ideas as it occurred between the natural history of plants and the language schemes.

The history of scientific taxonomy in general and plant taxonomy in particular follows our schema of decontextualization, objectification, abstraction, analysis and arrangement. Immediately following Aristotle's elaboration of the first principles of natural things, his disciple, Theophrastus, applied them to plants. The essential characteristics (differences) which he used for classification were chosen on logical and philosophical grounds, i.e. they were based on the function of the morphological structure in the life of the plant. As Aristotle understood, analyzed and arranged animals as decontextualized, abstract entities (structures, forms) so Theophrastus understood and analyzed plants.

The theoretical research of Aristotle and Theophrastus was followed by the practical studies of Dioscorides, who, in the first century A.D., wrote *De materia medica*, which treated plants as an adjunct of medicine, that is, they were considered as pharmaceutical 'simples' and were analyzed in terms of their medicinal qualities, powers, virtues and uses.[28] Descriptions of plants in Dioscorides' treatise provided information about plant features, e.g. shape, color and their habitats, which was meant to aid identification and location. The main part of the description was taken up with how the plant could be used. This is very close to (oral) folk wisdom and folk classification and very remote from logical analysis. It is folk wisdom put into the form of a 'how to' book.

It should also be pointed out that neither the theorists, Aristotle and Theophrastus, nor the practical Dioscorides were 'primarily interested in collection, description and classification as ends in themselves. Aristotle, the zoologist, and Theophrastus, the botanist, were always philosophers – their purpose to investigate the functioning of living organisms; Dioscorides studied botany as the servant of medicine.'[29] It may be the case that the relatively small number of species known to them – 'neither Aristotle nor Theophrastus knew more than about five hundred distinct kinds of plants or animals' – did not make precise definition and classification crucial.[30] As the number of specimens grew and as knowledge became more specialized, so too did taxonomy.

Until late in the sixteenth century, the *theoretical* study of nature, based on philosophical principles and aiming at systematic classification, had a tenuous history. It was all but dead until the rediscovery of Aristotelian texts in the thirteenth century when it was resuscitated by Albertus Magnus. His work however aroused little interest and until the late sixteenth century theoretical botany was written only sporadically, in scholarly quarters.[31] To the extent that the idea of a system of nature survived, it survived in the tradition of the Great Chain but there was little detailed arrangement of the species. Nor, furthermore, was the classification, such as it was, made in accordance with rigorous, motivated criteria.

We see this survival of informal Aristotelian classification in the great medieval compilations. *De proprietatibus rerum* by Bartholomaeus Anglicus (thirteenth century) presents the following classification of subjects: (1) God (2) angels (3) the soul (4) the elements and humors of the body (5) the parts of the body (6) the general properties of man: age, sex, etc. (7) infirmities and remedies (8) the world and the heavenly bodies (9) time (10) the elements (11) the air and weather (12) birds (13) water and fish (14) the earth and its physical features (15) the provinces (16) stones and metals (17) herbs and plants (18) animals (19) the accidents: color, taste, odors, etc. Not surprisingly the chapters on natural history were the most numerous: 39 on birds; 26 on fish; 197 on herbs and plants; 117 on animals.[32]

By 1491, a comparable book, *Hortus sanitatis*, shows increasing specialization with 530 chapters on herbs, 164 on animals, 122 on birds, 106 on fish and 144 on stones.[33]

During the Renaissance, contrary to what we might expect, most natural histories and botanical studies were essentially medieval. Those that were not were the exception. This means that they retained the qualities which linked medieval texts with an oral culture. In the perception of natural phenomena, for example, decontextualization was not extensive; abstraction was minimal.

The discussions are often a curious blend of fact, legend and popular belief, which embed the object in layers of 'life' – where life is the context of practice, tradition and belief, all that has been transmitted about or said of something whether in writing or in speech. While they are admittedly an extreme case, the works of Aldrovandi stand as a testament to this conception of natural history as hearsay and tradition. 'On Chickens', the fourteenth book of his *Ornithologiae* (1600) is typical of his method. The discussion of the particular kinds is preceded by a generalized treatment of the subject, 'Concerning Domestic Fowl Who Bathe in the Dust – The Chicken, Male and Female'. The topics included in this chapter are:

> Terminology; Synonyms; Differentiae of the Genus; The Form and Description of the Rooster and the Hen in Their Genus; Anatomical Details; Sex; Sight and Taste; Voice and Song; Lustfulness, Coition, Parturition, Incubation, Generation, Egg-Laying; Rearing and Feeding; Nature, Habits and Character; Magnanimity and Fighting; Sympathy and Antipathy; The Diseases of Chickens; The Method of Catching Chickens; History; Presages; Derivatives; Use in the Sacred Rites of the Pagans; Auguries and Prodigies; Mystical Interpretations; Moral Interpretations; Secret Signs (Hieroglyphics); Dreams; Emblems; Riddles; Epitaphs; Apophthegms; Proverbs; Fables; Apologues; Use in Medicine; Injurious Effects of Chickens; Use of the Chicken as Food; Various Other Uses of the Chicken; Insignia, Images and Coins.[34]

These medieval qualities were reinforced rather than diminished by the invention of printing. The initial stages of printing were devoted almost entirely to the mechanical reproduction of medieval manuscripts rather than to printing original works, and the reproduction of medieval natural histories continued through the sixteenth century. The press and its products were therefore at first a mechanical extension of medieval scribal culture with its close affinities with folk culture. Printing brought medieval culture into the hands of a significantly large number of people. As a result of this, in the sixteenth century medieval knowledge did not diminish: rather it increased. 'Many medieval world pictures were duplicated more rapidly during the first century of printing than they had been during the so-called Middle Ages. They did not merely survive among conservative Elizabethans...They became more available...in the 16th century than they had been [in] the 13th century.'[35]

Two good examples of this are the *Herbarium* of Apuleius and *The Book of Secrets* of Albertus Magnus which enjoyed great popularity in the sixteenth century. In their particular descriptions, natural objects are understood primarily by traditional use and hearsay value. We can see this in the description of diamonds in *The Book of Secrets:*

> If thou wilt overcome thy enemies. Take the stone which is called *Adamas,* in English speech a Diamond, and it is of shining colour, and very hard, in so much that it cannot be broken, but by the blood of a Goat, and it groweth in Arabia, or in Cyprus. And if it be bounden to the left side, it is good against enemies, madness, wild beasts, venomous beasts, and cruel men, and against chiding and brawling, and against venom, and invasion of fantasies. And some call it *Diamas.*[36]

It was paradoxically this medieval quality of the Renaissance natural histories that accounts for the persistence of the Great Chain of Being and the folk taxonomy it provided and/or reinforced. It was not therefore the Elizabethans who invented their world picture but rather that the invention of the printing press popularized for the first time on a large scale the essentially medieval conception of the Great Chain of Being.

The situation was not unlike that surrounding the rise of the novel in the eighteenth century: the emergent middle class demanded, funded and read books that mirrored their lives. Renaissance man required and found in his books the embodiment of his fundamental connection with his medieval predecessors and their world view. It was not until the end of the sixteenth century that radical changes in this medieval paradigm began to occur.

When we turn to the study of plants in particular, we can trace a similar evolution from a conception of nature as speech or discourse or life to a conception of nature as decontextualized items and their elements. The majority of plant studies in the Middle Ages and through the sixteenth century were in the Dioscoridian (herbal) tradition where the primary consideration is the medicinal qualities and uses of the plant. While identification of specimens was important, as it always is, the arrangement of these into an explicit system was not of particularly great concern. Witness for example this classification in the *Grete Herbal* of 1526: 'Fungi ben mussherons... There be two maners of them, one maner is deedly and sleet them that eateth of them and be called tode stoles, and the other dooth not.'[37] Rather than classification, what a typical treatment of a plant in this tradition might include is brief bits of information about its various names, where it grows, when it grows, how it looks; there would be more lengthy discussions of its effect on the body and, possibly, recipes for its compounding and directions for its application. A typical example is found in the popular Banckes' *Herbal* of 1525:

Asterion or Lunary groweth among stones and in high places. This herb showeth by night. This herb hath yellow flowers whole and round as a cockbell, or else like to foxgloves. The leaves of this herb be round and blue, and they have the mark of the moon in the middle, as it were three-leaved grass, but the leaves thereof be more [larger] and they be round as a penny; and the stalk of this herb is red, and this herb seemeth as it were musk, and the juice thereof is yellow. And this herb groweth in the new moon without leaves, and every day springeth a new leaf to the end of fifteen days, and after fifteen days it loseth every day a leaf as the moon waneth, and it springeth and waneth as doth the moon; and where it groweth, there groweth great quantity. The virtue of this herb is thus. They that eat of the berries or of the herb in waning of the moon, when it is *in signo virginis*, if he have the falling evil, he shall be whole thereof; or if he bear this about his neck he shall be helped without doubt.[38]

By the end of the century, attention focuses more on observing and describing the visual physical characteristics, particularly the structure. The plant is seen more as an abstract, isolated object. We see this in Gerard's *Herbal* (1597) where the typical herbal information is given, but it is laid out in a systematic way. The description, place, time and virtues are sections literally blocked out on the page. The descriptions themselves tend to render the plants with a greater degree of detail and precision, as can be seen in the description of the daffodil:

The first of the Daffodils is that with the purple crowne or circle, having small narrow leaves, thicke fat, and full of slimie juice; among the which riseth up a naked stalke smooth and hollow, of a foot high, bearing at the top a faire milke white floure growing forth of a hood or thin filme such as the flours of onions are wrapped in: in the midst of which floure is a round circle or small coronet of a yellowish colour, purfled or bordered about the edge of the said ring or circle with a pleasant purple colour; which being past, there followeth a thicke knot or button, wherein is contained blacke round seed. The root is white, bulbous or onion-fashion.[39]

An interesting side development in the Dioscoridian tradition is the doctrine of signatures developed by Paracelsus. According to Paracelsus bodies were composed of three principal elements: sulphur, salt and mercury. The essence and hence specific virtues of plants are determined by the mix of these visible elements. In order for us to know these (invisible) virtues, the

Creator has stamped plants with signatures or visible characters of their virtues and uses. What is interesting here is that there is a link between the visible characteristics and the virtues and essence. 'I have oft-times declared', says a seventeenth-century Paracelsian, 'how by the outward shapes and qualities of things we may know their inward Vertues, which God hath put in them for the good of man.'[40] Porta's *Phytognomonica* (1588) gives some examples: things with 'scaly surfaces like pinecones, thistles, catkins, and the overlapping skin on lily bulbs were the appropriate remedies for scaly conditions of the skin'.[41] Herbs with yellow sap cure jaundice; maidenhair is a cure for baldness.[42] William Coles' *Adam in Eden* (1657) draws the Paracelsian similarities between a pinecone and teeth: 'Crollius in his book of Signatures saith that the woody scales, whereof the Pine Apple is composed.. do very much resemble the formost teeth of Man; and therefore Pine leaves boiled in Vinegar make a good decoction to gargle the mouth for asswaging immoderate pains in the teeth and gums...'[43]

In the mainstream, we find in the evolution of the herbals an increase in observation and with this goes the tendency to devote more attention to the visual appearance of the plant. There is also an increasing separation of plant descriptions from a living context – from legends on the one hand and medicinal considerations on the other. There is a narrowing of focus to description of the plant structure itself: its tell-tale signs, its features, size, shape, color, etc. Increasingly, plants are treated as objects which are seen and observed rather than as participants sharing in life experiences.

With the work of Caspar Bauhin, the process of decontextualization seems to be complete. Bauhin's *Pinax theatri botanici* (1623) was the major reference work until the end of the century, identifying 6000 specimens (compare with the 500 listed in Fuchs' *De historia stirpium* of 1542) and classifying them on the basis of physical features alone. This had been preceded by a volume of illustrated plant descriptions (*Prodomus theatri botanici*, 1620) in which discussion of virtues and uses is dismissed as non-essential. The descriptions themselves are precise and concise observations of the plant as object, even more specifically as a form or structure, as can be seen in the entry for beet:

> From a short tapering root, by no means fibrous, spring several stalks about 18 inches long: they straggle over the ground, and are cylindrical in shape and furrowed, becoming gradually white near the root with a slight coating of down, and spreading out into little sprays. The plant has but few leaves, similar to those of *Beta nigra*, except that they are smaller, and supplied with long petoiles. The flowers are small, and of greenish yellow. The fruits one can see growing in large numbers close by the root, and from that point they spread along the stalk, at almost every leaf. They are rough and tubercled and separate into three reflexed points. In their cavity, one grain of the

shape of an Adonis seed is contained; it is slightly rounded and ends in a point, and is covered with a double layer of reddish membrane, the inner one enclosing a white, farinaceous core.[44]

It is interesting to compare these morphological descriptions with those formulated by Linnaeus whose works we may say carry the processes of decontextualization and classification to their ultimate point of development. In the example below, the plant is not described in running, connected discourse but is analyzed as a set of variables that are connected only by the 'syntax' of the position on the page:

Urtica

Masculi flores

CAL. *Perianthium* tetraphyllum: *foliolis* subrotundis, concavis, obtusis.

COR. *Petala* nulla.

Nectarium in centro floris, urceolatum, integrum, inferne angustius, minimum.

STAM. *Filamenta* quatuor, subulata, longitudine calycis, patentia, intra singulum folium calycinum singula. *Antherae* biloculares.[45]

It would seem to be the case that this analysis of the organism into an abstract structure with variable parts or elements, a decontextualized thing, is the condition necessary to proceed with arrangement and specialized classification.

It is also worth noting that, as verbal description developed, so too did pictorial representation. In a study which traces the evolution of herbal drawings, we see that in the fifteenth century new woodcuts were cut from copies of older blocks, with a resulting loss of detail and accuracy. One of the best examples of these bastardized woodcuts is to be seen in the *Grete Herbal* of 1526 (whose descriptions are of roughly the same quality as those in Banckes' *Herbal* cited above). The woodcuts made for Brunfels' *Herbarium* in 1530, however, take plants themselves rather than pictures as their model and there is a remarkable increase in detail. There is so much detail in fact that these illustrations have been criticized 'as being portraits of particular plants, showing not only their personal forms and characters but the very accidents of their growth, such as wilted leaves and broken stems, rather than being schematic statements of the distinguishing characteristics of the species and genera'.[46] This situation was remedied in the illustrations of Fuchs' herbal where author and artist collaborated to present 'careful schematic representations of what were considered the generic forms'.[47] This is perhaps the clearest example of the tendency towards abstraction that accompanies decontextualization.

iv. Botanical taxonomy in the sixteenth and seventeenth centuries

The botanical tradition inherited by the Renaissance was a fully developed folk taxonomy and as such fully subject to the universal laws which folk taxonomies follow. Ethnobiological investigations conducted on 'primitive' or developing societies today are equally pertinent to the 'primitive' or developing society of Europe. Those ethnobiological studies indicate that folk cultures universally divide natural entities, i.e. the plant and animal kingdoms, into five or six hierarchical levels which are reflected in the terminology and vocabulary of the language. These studies assert that the following levels of taxa are found (we will confine ourselves to examples from plants):

1. Unique Beginner: e.g. plant, animal.
2. Life Form: e.g. tree, bush, flower, weed, fern.
3. Intermediate: this is an unstable category that manifests itself during a period of adjustment in the taxonomic system and then disappears when a settled, adjusted system is re-established.
4. Genus: pine, oak, masterwort.
5. Species: ponderosa pine, black oak.
6. Variety: northern ponderosa pine, swamp white oak.[48]

What began to evolve somewhere around the mid-sixteenth century, and clearly by the seventeenth century, was an elaboration and specialization of this basic paradigm. By the early seventeenth century, naturalists and particularly botanists were engaged in what we may call 'specialized', scientific classification. The first shape this new taxonomy takes is a 'deepening' of the hierarchy, viz. there is an emergence of taxa at the level between life form and generic. We will call this level or these taxa *specialized.* Although they will not concern us as much, it is the taxa beneath the level of variety that concern specialized taxonomy in its more advanced stage. Comparison of Berlin's folk taxonomy with a fully developed hierarchy of specialized taxa will indicate the areas we mean:

Folk taxonomy	*Scientific taxonomy*	
Unique Beginner	Kingdom	
Life Form	Phylum	
	Class	Specialized taxa
	Order	Specialized taxa
	Family	Specialized taxa
(Intermediate)		
Genus	Genus	
Species	Species	
	Sub-species	
Variety	Variety	

The reasons for the elaboration or deepening of the hierarchy are many, but they are not unrelated. They can perhaps be subsumed under the general heading 'increase of knowledge' or, to use a phrase familiar in the seventeenth century, the 'advancement of learning'. We will discuss this in more detail in the next chapter but, briefly, that 'increase' was an increase in knowledge of natural specimens, particularly plant specimens. It came from two sources: Renaissance explorations of nature, at home and abroad; and information about plant specimens transmitted through recently translated, published, and disseminated classical texts. Quite simply, there were more pieces (of things, of information) and a greater variety among them. Description and identification of the new things came first; but these were necessarily accompanied by classification. A kind of classification more specialized and more detailed than folk classification was necessary if for no other reason that 'some sort of order was necessary for convenience in dealing with a large number of kinds'.[49] The Great Chain of Being simply did not have enough links to accommodate this new information.

If a traditional paradigm – and the taxonomy of the Great Chain was *the* traditional paradigm – is upset because of an information explosion, there will be a counterbalancing movement towards imposing order: in this case, the order of scientific, specialized taxonomy. (The development of mathematics is an imposition of a different kind of specialized order.) We simply see over and over again in the history of man and in all of his multitudinous endeavors, from the arrangement of his furniture to the arrangement of his ideas, that the drive to impose order on chaos inevitably and invariably emerges. The natural histories, botanical taxonomies and artificial languages are manifestations of such an emergence.

Turning specifically to botanical texts of the late sixteenth century, herbalists, now more specialized than before, began to drop the old way of alphabetically arranging plants; among other reasons, many new specimens were previously unknown and hence unnamed, making alphabetical arrangement impossible. Also, Theophrastus had been translated in 1483 and as his works became known among (specialized) herbalists, so did his rudimentary notions about systematic classification.[50] Furthermore, as herbalists became more familiar with the vegetable world through the new range of specimens, they made intuitive groupings based on the virtues of plants; of, what seemed to them, kinds existing in nature. In general we can say that in the last half of the sixteenth century and in the early seventeenth century increased observation and knowledge, combined with decontextualization fostered by an increasingly literate culture, brought increased discrimination. Increased discrimination is, at a certain point, chaotic. Out of this chaos came increased classification.

It was intuitively felt by the sixteenth- and seventeenth-century herbalists that new taxonomic levels were necessary due to inconsistencies created in the received paradigm of folk taxa by new specimens. While this need was felt, it was by no means clear how, and according to what criteria, the new groupings could or should be made. The history of botany between 1550 and 1650 is a process through which this intuition was articulated and a theory of classification was developed.

If we look at the *Crŭÿdeboeck* of the herbalist Dodoens (1554) we can see the basic outlines of the taxonomic confusion.

> Book I...lists herbaceous plants in alphabetical order. Book II is on flowers used in garlands and bouquets, or for their aroma, and also treats of the umbellifers. Book III is on roots, vines, poisonous plants, and cryptogams. Book IV concerns cereals, leguminous plants, and those of the marsh and seashore. Book V treats of edible plants, and Book VI is about shrubs, trees, forest trees, and evergreens.[51]

We should begin first by pointing out that the category of unique beginner is quite fixed and never enters into botanical discussions. The categories of life form and genus are present in folk knowledge and in the vernacular. Dodoens' classification of plants starts at a level above genera, i.e. it is a level of taxa which groups genera. In his classification, however, we see a combination of items and a mixture of taxa, i.e. there are those that would normally be considered life forms, like 'herbs', 'trees', 'shrubs', 'flowers', 'vines'; but we see also those that are more specific than life forms, e.g. 'evergreens', 'cereals', 'edible plants', 'poisonous plants', etc. We can predict that this classification which straddles the level of life form and something subordinate to it (but higher than genus) is due for a split. Furthermore, it is clear that a number of criteria (use, virtue, location, structure, etc.) rather than one essential difference are used to distinguish the six plant categories Dodoens creates. There is no conscious consideration of criteria; there is no theory.

Caspar Bauhin's *Pinax theatri botanici* (1622) is the most complete of the herbals with about 6000 species listed, the largest number to be treated by the herbalists. In his classification Bauhin consciously invokes Theophrastian principles to make a clear distinction between herbs (Books I–X) and trees and shrubs (Books XI–XII). Both are life forms. The herbs are presented in ten books or specialized classes which are to some extent based on structure. These are arranged according to Aristotelian degrees of perfection, from the perfect (having a seed which is related to the plant's primary function of reproduction) to the imperfect (having no seed as was believed to be the case with ferns and mosses). Each of Bauhin's books contains six sections which in turn contain a number of classes meant to be genera; under each genus the species are listed.[52]

What we see in Bauhin's work is the need for a specialized level of taxa but this level has not yet been clearly articulated. Rather, it is distributed among the 'books' and 'sections' of the outline.[53] An interesting indication of this ambiguity is found in the fact that only some of the classes are named and then in an irregular way: Book II is entitled 'de Bulbosis', Book III, 'Olera et Oleracea'. And while most of the sections are left unnamed, Book VI, section iv is entitled 'Umbelliferae', Book VIII, section i, 'de Convovulis'.

John Parkinson's herbal, *Theatrum botanicum* (1640), has been called 'the last of its kind'.[54] It represents not so much a development as a summation of the herbalist tradition.

Parkinson's 17 major classes, which he calls tribes, are given below:

1. Plantae Odoratae. Sweete smelling Plants; 2. Catharticae sive Purgantes. Purging Plants; 3. Venenosae, Narcoticae, Nocivae & Alexipharmacae. Venemous, Sleepy, and Hurtfull Plants and their Counterpoysons; 4. Saxifragae Plantae, Nephriticae sive Calculum frangentes. Saxifrages, or Breakstone Plants; 5. Plantae Vulnerariae & Ferruminantes, id est, Consolidantes. Vulnerary or Wound Herbes; 6. Plantae Refrigerantes & Intubaceae. Cooling and Succory-like Herbes; 7. Plantae Calidae & Acres. Hot and sharpe biting Plants; 8. Umbelliferae. Umbelliferous Plants; 9. Cardui & Spinosae Plantae. Thistles and Thorny Plants; 10. Filices & Herbae Capillares. Fearnes and Capillary Herbes; 11. Legumina. Pulses; 12. Cerealia. Cornes; 13. Gramina, Junci & Arundines. Grasses, Rushes and Reedes; 14. Plantae Paludosae Aquaticae & Marinae, Musci & Fungi. Marsh, Water and Sea Plants, and Mosses, and Mushromes; 15. Miscelanea. The Unordered Tribe; 16. Arbores & Frutices. Trees and Shrubbes; 17. Exoticae & Peregrinae Plantae. Strange and Outlandish Plants.[55]

In his table, we see a division between herbs and trees and shrubs, and amongst the herbs a tendency of arrangement according to the perfection of the plant. The major taxa show the usual combination of life forms (trees, shrubs, ferns, grass) and specialized taxa (umbellifers, saxifrages) although as was the case with Bauhin there is more attempt at specialization.

The herbalists do not tell the whole story however. With the increasing specialization of knowledge also went the attempt to order by theory rather than intuition. It was therefore the philosopher and theoretician Andrea Cesalpino who re-established Aristotelian principles and set up a groundwork for systematic classification. Cesalpino's work in *Questionum peripateticorum* (1571) and *De plantis libri XVI* (1583) is the first (modern) example of Aristotelian theory being applied to the classification of natural things. Although his work was little known in his own day, he was the acknowledged master of all seventeenth-century taxonomers and even of Linnaeus. To the extent that his ideas became orthodoxy, so too did his explicit Aristotelian essentialism.

Cesalpino was thoroughly versed in the works of Aristotle (as distinct from the works of his commentators) and adopted Aristotle's theories that science is the definition of essences: 'The determination of a concept according to the next higher genus and specific differences reproduces the process by which substance unfolds itself in special forms of being; a complete system of scientific definitions, then, is a complete expression of the substantial forces controlling reality.'[56]

The means to this end is botanical classification which 'like all science is the rapprochement of like, the distinction of unlike, accomplished through distribution in genera and species, according to the differentiae translating the essences of things'.[57] 'We seek out similarities and dissimilarities of form, in which the essence of the plants consists...'[58]

The problem in the classification of plants (or anything for that matter) was determining what the differences should be. The herbalists, as we have seen, used different kinds of criteria – plant morphology, habitat, medicinal use – as it pleased them. Or they used several of these characters together in determining any particular species. Realizing that classification must proceed by a single character, a feature which defines the form or essence, i.e. a feature that is essential to the nature of the plant, Cesalpino set out to determine logically which kind of characters those should be. He based his rationale for character choice on Aristotle's theory that it is in the nature of plants to have a nutritive soul and that the end of nutrition is reproduction. 'Since the function of the nutritive soul consists in producing something like itself and this like has its origin in the food for maintaining the life of the individual, or in the seed for continuing the species, perfect plants have at most two parts...one part is called the root by which they procure food; the other by which they bear the fruit, a kind of foetus for the continuation of the species...'[59] On the basis of nutritive function, Cesalpino defines the stem, a structural character, as the essential part of the plant. The first division in his taxonomy is therefore according to hard (woody) stems and soft (herbaceous) stems.

Cesalpino continues this line of reasoning by arguing that since the next activity of the plant is reproduction, subsequent divisions should be made according to the organs of fructification. Hence he divides plants according to the number, position and form of the reproductive parts, the structure and number of seeds and the structure of parts containing the seed. Finally, he comments that 'Other marks to be derived from roots, stems and leaves may

may be used...for forming the smaller divisions.'[60] He concludes by rejecting those 'marks which contribute to the constitution neither of the whole plant nor of the fruit, such as colour, smell, taste' as mere accidents.[61] With this, Cesalpino dismisses the practice of the herbalists to classify on the basis of plant 'habit' (general disposition) and institutes instead structure as the essential character signifying the form which is connected to the essential activities of the life of the plant. Structure becomes the exclusive criterion for study and classification.

This reduction of plants to 'structure' in a taxonomy, which is itself a structure, is not a coincidence; the coming of the two together marks the final point in the process of abstraction and decontextualization. The concept of structure also establishes the principles by which individuals can be analyzed, and genera and species be determined. Finally, by means of the criterion of structure, Cesalpino was able to establish a new level of taxa (itself an abstraction) which we call 'specialized'. Cesalpino's work established logical classification and defining as rational, specialized, scientific activities.

As would be expected from a theoretician, we see in Cesalpino's classification the emergence of the specialized taxa as clear-cut, well-defined, distinguishable forms, determined by consciously chosen criteria and systematically given names. The folk-level life form is accommodated by the classes *Arboreae* and *Herbaceae*. Then on a separate level are the specialized taxa:

Life form	Herbaceae	
Specialized	Solitariis Seminibus	
Genus		Semina nuda, papposa: Valeriana
		Semina pericarpio obducta: Daphne, Jasminum
		Flos in summo fructus: Osyris, Valerianella
		Flos in sede fructus, semen calyce exceptum: Urtica...
Specialized	Solitariis Pericarpiis	
		Flos exterius situs (Pomum): Curcurbitaceae
		Flos inferius situs (Bacca): Solanaceae...
Specialized	Solitariis Vasculis	
		Legumina: leguminous herbs
		Capsulae: Caryophyllaceae...[62]

The next theoretical botanist was Joachim Jung, a seventeenth-century polymath who had many connections with the artificial language projectors. His reputation was based primarily on his work in mathematics and chemistry, but he was also known for his popular Hamburg logic, and, to a small circle, for his botanical accomplishments. Jung studied medicine at Padua where he became familiar with the botanical theories of Cesalpino. His botanical work was carried on throughout his life, but his only botanical writings were published posthumously: *De plantis doxoscopiae physicae minores* was published in 1662 followed by his *Isagoge phytoscopia* in 1678. His notes however circulated for some time before this.[63]

Jung was not unknown to the language projectors. In his early days, Jung was in the circle gathered around Wolfgang Rathke, a Baconian educational reformer and teacher of Comenius. Around 1619–20 he associated with various Rosicrucians and was suspected of belonging to the sect, as were many others in Comenius' circle. Among his Rosicrucian colleagues was John Valentin Andreae, author of the utopia *Christianopolis* (1619) and one of the mentors of Hartlib and Comenius. In the 1630s Jung was known in England through the Hartlib connection. In 1642 Comenius met with him in Hamburg and shortly thereafter tried (unsuccessfully) to enlist Jung's help in editing his works. From 1644 to 1645 Sir Charles Cavendish studied mathematics with Jung and sent outlines of a now lost mathematical treatise back to London. In 1647 Jung was in correspondence with Henry Oldenburg, Hartlib's father-in-law and the first Secretary of the Royal Society. In the same year, Cyprian Kinner indicated that he expected Jung to help with tracing the taxonomic features of things for the educational treatise that succeeded his proposal for a universal language. We do not hear of Jung again until 1654 when Hartlib sent Boyle a copy of Jung's *Protonoeticae philosophiae sciagraphia*. Comenius was again meeting with Jung in 1656; Jung died the following year.

In his botanical works, Jung follows the lead of Cesalpino in applying Aristotelian principles to the analysis of plant life. He followed Cesalpino's Aristotelianism also in taking structure as the fundamental criterion for plant analysis and carried this one step further by providing a detailed system of plant morphology as well as a basic set of terms and definitions for morphological analysis, an inevitable development from the increasing tendency to treat plants as abstract structures.

His outline of plant morphology was achieved by applying logical division and the result is a dichotomous taxonomy of morphological units, mutually exclusive, determined by single characters.[64] Perfect plants have two parts: a subterranean part (root) for nourishment and an upper part (stem) for propagation. The upper part consists of stalk, leaf, flower, fruit. Each of these is defined and analyzed into component parts. The parts of flowers, for example, are stamen, stylus and folium (corolla).[65]

Although Jung devoted most of his attention to plant morphology, he also constructed a brief taxonomical outline. He objected to Cesalpino's major divisions of plants into trees and herbs on the logical grounds that the division did not hold good in all instances and the categories were ill-defined.[66] He proposed instead his own logical division based on plant morphology:

I. Perfect plants, possessing both stem and root
  A. Plants with distinct stem and leaves
  B. With a structure combining the natures of stem and leaf
II. Imperfect plants, lacking either stem or root:
  A. Without a stem; putative seeds borne on the backs of the leaves
  B. Without a root
    1. With distinctive stem and leaves
    2. With a stem only
    3. With a structure combining the natures of stem and leaf.[67]

The next taxonomer, John Ray, leads to the language projectors, since it was Ray who composed the tables of plants for Wilkins' *Essay*. Here we need to look at Ray's contributions to botany prior to his collaboration. This brings the history of taxonomy up to the time of the universal languages.

It is interesting to follow Ray's taxonomic work from its beginning in 1660 to its end after the turn of the century (it takes us almost up to the time of Linnaeus). What we see in that development is the entire history of taxonomy: Ray's early works show a blend of the herbal and Aristotelian traditions; his middle work, including that which he did for Wilkins, is theoretical and Aristotelian; his final work repudiates Aristotelian theory just as Ray himself repudiated Wilkins' philosophical language. We shall discuss his Aristotelian taxonomy when we treat Wilkins, and his final position when we consider the demise of essentialism and the universal language movement. At this point, we shall look at his early work: emerging from the herbal tradition, attempting to deal with new specialization of information, grasping for principles which would impose order.

Ray began his career as a Fellow of Trinity College, Cambridge and it was here that he met Wilkins when, in 1659, Wilkins was appointed Master. Although Ray held Aristotelian visions of a natural system which he believed it was the naturalist's work to uncover, his first botanical publication *Catalogus plantarum circa Cantabrigiam nascentium* (1660), was primarily descriptive with an alphabetical arrangement. At the end of the catalogue, he briefly sketched an outline of a plant taxonomy which, as he himself admits, simply follows 'the more usual headings' of Bauhin and other herbalists. He was familiar with Jung's *Isagoge* which he had received in manuscript.[68] Ray starts his classification with the superordinate category, perfect/imperfect. Then, in the tradition of Bauhin and Cesalpino, we see two clearly defined taxonomic levels: one corresponding to life form (composed of trees, shrubs, sub-shrubs and herbs); the other, the new specialized taxa. In the classification, he uses various criteria rather than a single character to determine the specialized taxa:

Life Form	IV. Herbs		
Specialized	1. Bulbosae	12. Rotundifoliae	
	2. Tuberosae	13. Nervifoliae	
	3. Umbelliferae	14. Stellatae	
	4. Verticillatae	15. Cerealia	
	5. Spicatae	16. Succulentae	
	6. Scandentes	17. Graminifoliae	
	7. Corymbiferae	18. [omitted in the original]	
	8. Pappiflorae	19. Oleraceae	
	9. Capitatae	20. Aquaticae	
	10. Campaniformes	21. Marinae	
	11. Coronariae	22. Saxatiles[69]	

Although structure plays no special part as a criterion, in his concluding remarks Ray shows awareness of Jung's theoretical principles of determination by morphological criteria exclusively. Ray indicates that plants can be arranged according to their roots, stalks, flowers, seeds or leaves, etc.[70] In his next taxonomical endeavor, the next step forward in the history of botanical taxonomy, he was finally to apply these structural principles.

Ray published this in 1660, the time when Dalgarno and Wilkins were discovering that a truly philosophical language needs a taxonomic base. It is not a coincidence that the next development in taxonomy came when Ray created the tables of plants in Wilkins' *Essay*. In their respective desires for order, they shared the same methodological and intellectual heritage of Aristotelianism. In order to impose order, they classified; in classifying, they had to establish a new specialized level of taxa; in order to establish these taxa they sought philosophical justification for the criteria of division; in order to find that justification, they turned to Aristotelianism; and behind all of this was decontextualization, the conception of things as abstract structures to be analyzed into their component parts. The next step was the 'naming of parts'.

As an actual written thing, rather than as a mental process of classifying, any taxonomy has two parts: in addition to the taxonomic structure there is also a set of taxonomic names or a nomenclature which expresses the taxonomic structure. These may be thought of as two sides to a coin, for the one does not occur without the other. The aim of the botanical classifiers – Cesalpino, Jung, Bauhin, Ray – was to discover the order of nature and create a systematic representation of it; their emphasis, we may say, was on the taxonomic structure. The language projectors also wished to create a systematic representation of the order of things but their emphasis was on the reverse side of the coin as it were: it was on the taxonomic nomenclature. It is simply a question of how you see the coin, of whether you identify it by the head or by the tail. In creating their taxonomies of plants, and in creating their taxonomies of all known things, the linguists and the botanists were playing with the same coin no matter which side they started with. And just as it took until the time of Linnaeus for the botanists to realize that their taxonomic work was also linguistic, as we shall show in subsequent chapters, so it took the linguists a while to understand that their linguistic work was taxonomic.

CHAPTER 3

Nomenclature

i. Taxonomy and stages of language development

Our minds contain the sum of all our individual knowledge; one kind of knowledge we have is a mental classification of things, a knowledge of categories, the features that account for identity and difference, and the relations between the categories, in part, if not in whole. One of the facts we know about a thing is its name. We can transcribe, represent, articulate, draw up and put down this knowledge in a variety of forms. An encyclopedia records all the facts about things known to a group of individuals; a taxonomic table records the features, categories and relations; a dictionary records words, included in which are the names for things, and the definitions of words. These representations of our mental contents are achieved through the medium of language; they are drawn up and put down in words, arranged or deposited in a marked-out space. The archetypal spatial repository of words is the book.

To the best of my knowledge there has never been an encyclopedia, taxonomic table, or dictionary set forth exclusively in pictures. This is probably because pictures and images, in one way or another, are particular and concrete. No matter how stylized or iconographic a drawing is, say, of a cherub it still portrays something that is a particular size (chubby) with a particular shape (baby-like). To represent abstract things directly, e.g. genera or species, which are not concrete individuals but classes of individuals, we need to use language. We can only represent the *category* flower or rose, in linguistic rather than pictorial images.

While it is obvious that an encyclopedia and a dictionary are books written in language, the rubric taxonomic 'table' might obscure the fact that it too is a linguistic representation; it is a representation of our mental order of things expressed through the medium of language, a discourse on the order of things. If we consider a taxonomy from this point of view, we see that it consists of a set of names for the categories of our mental taxonomy. The taxonomic structure we can think of as conceptual; it consists of our knowledge or recognition of categories and relations of identity and difference. The nomenclature is obviously linguistic; it includes a set of names through which the categories and their relations are expressed.

The relation between the taxonomic structure and the taxonomic nomenclature is the subject of this chapter. What we see is that in the natural language of the seventeenth century this relation is found to be unsatisfactory – to various specialists, to the intelligentsia, to scientists and to projectors alike. Their language activities, which are necessarily taxonomic activities, and vice versa, constitute an attempt to relate ideally structure and nomenclature. Simply put, this ideal relation provides a set of names isomorphic with the taxonomic structure. Every recognized category and every recognized member of the category, i.e. every space on the table, is given a unique designation. At the same time, this unique designation is constrained so that (1) there are no synonyms (more than one name per class or member); (2) there is no polysemy (more than one class or member per name); and (3) there are no unnamed classes or members. In addition to unique designation, the ideal set of names must preserve or express the structural relations of identity and difference.

We have already seen that a taxonomic structure conceived along Aristotelian lines presents us with an arrangement of the categories of things. It also states (and this is what makes it a unique kind of taxonomic structure) the essential, defining features of the categories. The two elements, category and difference, provide, in Aristotelian terms, a definition of a thing. This Aristotelian kind of taxonomy was adopted by botanists like Ray and by projectors like Dalgarno and Wilkins. Linguistically speaking, this kind of taxonomic structure results in a unique kind of naming, viz. the definitions of things act as the names for things. There is no separation between naming and defining in a taxonomy of this Aristotelian sort, just as there is no separation between naming and structure in any kind of taxonomy. The nomenclature proposed along Aristotelian lines provides unique designation, a statement of structural relations and a definition. To the extent that the taxonomic structure includes definitions, and definitions constitute names, Aristotelian taxonomic naming reflects taxonomic structure in the most transparent manner possible. The taxonomy is an encyclopedia, nomenclature and dictionary all at the same time.

By virtue of the fact that taxonomic tables classify, they define; by virtue of the fact that they define, they name; by virtue of the fact that they name, they can be conceived of as a kind of language. Language here means nomenclature, a set of names for classified items. More precisely, we can adopt Foucault's term for this: a set of essential nominations.[1]

With regard to the artificial languages, we should note that projectors like Dalgarno and Wilkins were in effect creating two artificial, philosophical languages. The first one was the 'language' of the taxonomic tables. The second one was the artificial language proper, represented in the newly constructed words; this one represented certain, essential features of the taxonomic nomenclature, but not all of them. If the taxonomic structure/nomenclature is a linguistic image of the order of things, the artificial language is an image of that image. The written character is an image of an image of an image.

The fundamental feature of the primary language, the taxonomic structure/nomenclature, is that the names for things are self-defining and transparent. This dimension is what distinguishes essential nominations from code names (which is what the artificial names were) and, for the most part, from the names of ordinary language.

There is an old question as to whether the chicken or the egg came first. There is a similar question as to whether categories exist prior to naming, or whether naming gives rise to categories. I have no way of knowing how conceptual categories develop, independently of language. We do have evidence, however, as to how taxonomic names develop. We argued in the last chapter that the seventeenth century saw a development in scientific taxonomy. Rather than attempt to discuss how taxonomic concepts and structures develop, perhaps universally, here we shall discuss how the taxonomic nomenclatures develop, perhaps universally.

As we indicated before, Berlin's studies assert that universally there are five or six levels of taxonomic structure which are manifested in the nomenclatures of ordinary language: universal beginner, life form, (intermediate), generic, specific and varietal. What Berlin's studies indicate in addition is that nomenclaturally there is a developmental sequence, i.e. that there is a life history to the way in which the vocabulary of a language develops relative to these categories.

> Generic names are fundamental and will occur first. These will be followed by the major life-form names and specific names. At yet a later period, intermediate taxa and varietal taxa will be labelled. Finally the last category to be lexically designated...of any ethnobotanical lexicon will be the unique beginner. The suggested sequence can be seen diagrammatically as follows:

$$\text{generic} \rightarrow \begin{Bmatrix}\text{life-form}\\ \text{specific}\end{Bmatrix} \rightarrow \begin{Bmatrix}\text{intermediate}\\ \text{varietal}\end{Bmatrix} \rightarrow \text{unique beginner}$$

...no temporal ordering is implied for some categories. Thus no claim is made as to the priority in time of specific names over major life-form names. On the other hand a claim is made that a language must have encoded *at least* one major life-form name and one specific name before the appearance of intermediate and varietal named taxa.[2]

Berlin had hoped to find that the development of specific names preceded the development of the more abstract life-form names, which intuitively seems to be the way it should happen. But he finds no solid evidence that this is the case, although he indicates that the evidence on this matter one way or another is very scanty.

Cultures differ as to the degree of development they have achieved relative to this developmental sequence; we will take this up in detail at a later point. Nevertheless, Berlin finds that *in general* 'most folk biotaxonomies are comprised primarily of named genera, major life-forms, and specific taxa, with generic classes being by far the most numerous and psychologically significant'.[3] The development of the categories is correlated with agricultural development, generic categories being universally basic; this can be explained perhaps by biology; perhaps by the fact that everyone will have some contact with the plant world in the form of food; perhaps by the fact that generic categorization appears to be 'part of' human thought.

The category of species is correlated with early cultural/technical development. Life form as a category seems to exist before it is named and life-form taxa evolve from items that are the most salient to the culture. The varietal category involves 'highly important groups of cultivated plants' and reflects 'greater control over the processes of domestication'. They are the product of 'relatively advanced horticulturalists'.[4]

Intermediate taxa tend to exist as covert categories that are not, or are rarely, named. When they are named, the names are short-lived. This suggests that there is little cultural need to label these categories. Berlin points out that little is known about these taxa, but he tentatively suggests that they are unstable categories which arise out of situations in which adjustments need to be made among the genera. Once the adjustments are made, the linguistic markers for the intermediate taxa tend to disappear. One frequent trigger of the adjustment is the introduction of new genera from foreign sources through culture contact situations.[5]

Names for unique beginners, e.g. 'plant', are relatively recent. As Berlin notes: 'In contemporary languages of primitive peoples, a single unique expression for "plant" is notably lacking and there is no reason to assume that such was not the case in historical times.'[6] Theophrastus for example created a name for plant by taking *phiton*, the everyday name for 'cultivated plant' or '(cultivated) tree'.[7] Ancient Latin similarly had no single word for the plant kingdom; *arbor* and *herba* were the most comprehensive class concepts (although this is not to say that they did not use one of these terms to refer to plants). According to Ullmann, 'the modern meaning of "plant" is first found in Albert Magnus in the 13th century, whereas the French *plante* did not acquire this wider sense until 300 years later'.[8]

It is perhaps obvious to say that names developed because men found new things they wished to talk about or new situations which required them to talk about new things. But that simply pushes the 'why' back one step further. What causes or contributes to the new situations? I would suggest that in the seventeenth century the relation between taxonomic structure and names in ordinary language was breaking down due to an information explosion aided and abetted by print/writing/literacy. And that it was breaking down in such a way that more advanced, specialized knowledge could not be accommodated in ordinary language. Just as there had not been a sufficient number of links in the Great Chain to accommodate more specialized taxa, so there were not a sufficient number of names to represent these new kinds of things. This required that a new language or nomenclature be created. This breaking down of the taxonomic nomenclature of ordinary language with the subsequent rise of a specialized language of taxonomy can be related to developments in literacy and changes in the media of communication. As Foucault has argued, classification was the episteme of the seventeenth century. The language which accompanies classification, the language of essential nomination, simply follows automatically.

Berlin's studies make no correlation between developments in writing/literacy and the developmental stages of folk taxonomy, yet to consider such development in that light is suggestive. The fact that life-form names, for example, develop simultaneously with species names is a case in point. We intuitively tend to think that abstraction comes at a time later than differentiation and specification. Perhaps if the dates for the introduction of writing were compared with the dates for the introduction of life-form names there might be a correlation; life-form names might be found to come after or with the introduction of writing.[9]

Similarly, varietal (and perhaps sub-varietal) names develop after life-form names but in primitive cultures they are not very extensive. In one case study, of 1094 terminal taxa, 961 (88%) were specific names and 97 (9%) were varietal names.[10] The reason for this could be attributed in part to the nature of communication media. In an oral situation, the number of names for terminal taxa (which should be the most numerous) might well be constrained by the limits of memory as well as of use. This would mean that varietal names would give way to the importance of specific names and specific names could not be developed beyond a certain point because of these constraints. Lévi-Strauss suggests that the maximum number of terminal taxa that can be handled orally is about 2000.[11] Writing lifts this mnemonic constraint and provides for the elaboration of an endless number of terminal classes. Thus with the introduction and spreading of written language we should expect varietal names to become more important and their numbers to grow. The number of all terminal taxa, including species, should, for that matter, grow. Such was the case during the seventeenth century. The herbalist, Bauhin, listed 6000 terminal taxa in 1623. By 1676, Ray had increased that number to 18,000. He had also begun investigating the precise, theoretical difference between species and variety.[12]

Intermediate names are a rather special case, but they too might be fruitfully considered from the perspective of language/communication changes. Intermediate categories are 'a response to situations whereby native polytypic generics must be distinguished from newly encountered generics'.[13] The newly encountered forms are first recognized to be similar to the native polytypic generics but there is also a recognition that they are not similar enough to be considered true species of the genus. Essentially what happens is that, despite the similarity, the new, foreign specimen is made a separate genus and a named, intermediate category is created to tie together or relate the foreign genus and the native genus. The similarity is expressed in the name of the genus and the name of the intermediate. The similar names eventually get lost, and the old genus and the new, foreign genus are considered to be two distinct forms. Once this linguistic similarity disappears, the need for the intermediate name also disappears. The similarity is still recognized conceptually, i.e. the intermediate category continues to exist covertly but it is not necessary to mark it linguistically. It is my feeling that the level of taxa I have called specialized originates in these intermediate, unstable categories.

One of the situations that leads to the creation of intermediate taxa is language and culture contact, which introduces new types which challenge the stability of the native configurations of genera. In the historical study of botany we see a number of situations that bear this out. The humanism of the Renaissance for example is a case of culture contact *par excellence*. Classical texts were 'recovered' and disseminated by printing/literacy, as was information about new Renaissance discoveries. In this case, we would expect that increase in culture contact and the spread of new, foreign information would create instability in the prevailing taxonomic and linguistic structure. We would expect therefore intermediate taxa, taxa formed to handle the instability, to appear. Such is the case with the specialized taxa. It remains only to be said that for a special group - botanists – the need for these taxa to be marked linguistically remains rather than disappears. And, in a fully literate culture, which meticulously preserves tradition, taxonomic names, once introduced, no matter how unstable, are preserved or 'fossilized' and do not easily disappear.

The appearance of the category of unique beginner can perhaps also be linked with writing. As many have pointed out, abstract discourse like philosophy, science and logic requires a written medium where terms are removed from the changeableness and flux of speech in order that they may be stabilized and scrutinized.[14] Significantly, as we see in the example of 'plant', names for unique beginners are found with the literate Greeks, and in the medieval scribal culture; they become fixed in the Renaissance, i.e. in the time of print culture when abstract, philosophical, scientific discourse flourished.

We should like therefore to suggest that the later stages of nomenclatural development can be seen as language changes, either brought about or aided by changes in the media of communication, i.e. further stages in taxonomic development and nomenclature co-occur with the introduction of writing/literacy. Finally, we have suggested in our discussion of classification that, once a taxonomy or nomenclature completes the four stages indicated by Berlin, it enters a fifth stage called specialization, a stage which is also marked linguistically and which is probably not possible without widespread use of written language.

The Prague School linguist, Havránek, has called this development the 'intellectualization' or 'rationalization' of the standard language, defining it as 'the adaptation of language to the goal of making precise and rigorous, if necessary, abstract statements'. It culminates in 'scientific or theoretical speech'. He identifies this specialization, therefore, as a stage of language development occurring with the rise of written language and with the development of science.[15] What this new stage requires of language are (1) unequivocable words (2) special distinctions in terms and (3) abstract summarizing terms.[16] The latter two are relevant to taxonomic levels in vocabulary; the first, which involves making definitions, creates the need for the taxonomizing in the first place.

The evolution of specialized nomenclature ties in with our contemporary understanding that the status of taxa is relative to the culture of the language group and to its various levels of specialization, division of labor, or expertise. As Berlin notes, in the developmental sequence he proposes, the category of genus is the universally basic one.[17] The 'universal' peoples and languages studied were primarily agricultural, in the sense that they were engaged in the cultivation of plants for their food. But this is not necessarily the case for urban people who live in a natural environment which is 'ever-shrinking (both literally and figuratively)'.[18] Urbanites, for example, may know virtually no specific names for plants or they may recognize linguistically generic names but have no knowledge of the organisms those names refer to. Other studies indicate that where *oak* is the basic (generic) term used by agricultural people, *tree* is the basic (generic) term used by some city folk.[19] Berlin generalizes that where loss or atrophy occurs, it will occur 'from the bottom up', i.e. the taxonomy will become more shallow.[20]

Contrarily, it may well be the case that scientific expertise causes the new taxonomy to become deeper, i.e. to add more levels and more specimens. This would appear to be the case with the development of the intermediate, specialized level which by the time of Linnaeus was expanded into three levels: family, order and class. It would appear also that the development of expertise would put more emphasis on the varietal level and would develop a sub-varietal level, as eventually was the case in post-Linnaean biology.

Furthermore, it would seem to be the case that atrophy and specialization are necessarily related. As roles fragment, as one group becomes expert in some area, it no doubt necessarily becomes ignorant in some other areas. We do not get renaissance men, except as exceptions, beyond the Renaissance.

As divisions of labor and differentiation of fields of knowledge occur, the language system is necessarily differentiated also. There is for some individuals an atrophy of taxonomic names in one domain with an increase in another. Taking all the idiolects together, the language gives rise to specialized terminologies and dialects, e.g. the dialect of botanical nomenclature. Seen from one perspective, the artificial languages were, in a sense, attempts to create specialized forms of the language they tried to bring up to date scientifically. Seen from another perspective, say the botanist Ray's perspective, they were not specialized enough. The very ambiguity of their degree of specialization marks them as products of a transition in developmental stages.

While Berlin limits his discussion to the naming of plants, his comments are pertinent to the general problems of naming in sixteenth- and seventeenth-century English. Until the early sixteenth century as a vernacular language, English was for the most part reflective of an oral culture (Chaucer et al. notwithstanding). From the early sixteenth century, English as well as other vernacular languages received more widespread attention than hitherto had been the case. This rise of English as a prestige language, as well as the rise of the standard dialect as the basis of that language (which can only have been established with writing and widespread literacy) has been extensively treated in the literature.[21] Without going into a complete analysis of the reasons for the rise in the importance of the vernacular and the problems of naming that accompanied it, we can note some of the significant features.

In the sixteenth and seventeenth centuries, vernacular languages assumed importance for a number of reasons: religious, political, social, economic, literary. Not the least of these reasons was the rapid spread of literacy made possible by the printing press. Many who were now able to read could do so only in the vernacular; they did not know, and could not read, Latin. There was a great demand by this constituency for the acquisition of knowledge; and there was a great desire to spread knowledge to them. There were only two linguistic alternatives here: teach them Latin, or transmit in the vernacular. As we will see with Comenius and his group, the latter was considered the more efficient.

Not only was the readership new: a good deal of the information to be communicated was new also. Sooner or later Latin would be found insufficient since there are few ways a dead (classical) language can grow with changing times. But given that one wanted to communicate specialized information in the vernacular, it too was inadequate. Most especially, it lacked the necessary terms and vocabulary, not only for the new but for the old as well. We see this in any number of examples. Those translating classical texts into English in many subjects, e.g. medicine, botany, mathematics, geography, religion, found that English lacked names for things discussed by the ancients. As with Latin, newly discovered phenomena lacked names also.

In providing new words and new names for the vernacular, there were two alternatives: borrow roots and whole words from Latin to the extent that that was possible, or create names from existing components of English by compounding. Each solution had its merits but each had its limitations. How to name was a widely and hotly debated issue.

In addition to the problem of creating new names, there was also the problem of defining. In order for new names to gain wide currency as well as precision and stability, they had to be defined. This is true whether the words were English-based or Latin-based. It was particularly true of the

Latin-based words used by vernacular speakers who knew no Latin. Furthermore, the vernacular languages were spoken languages, as opposed to Latin which was a print-governed language. The vernaculars were, therefore, subject to change, variety and fluctuation to an extent that Latin was not. Again, definitions and rules were felt necessary in order to fix the undesirable instability in vernacular language. These combined features gave rise to the large number of dictionaries produced during the seventeenth century.[22] These included lexicons which explained borrowed terms, terms of art, and hard words. Some people called for academies to perform this fixing function; others called for acts of Parliament and royal edicts.

One interesting proposal, not unlike those made in the artificial languages, solved the problem of creating new words and defining them simultaneously. It argued that new terms composed of English words automatically created terms that were self-explanatory and self-defining. Suggestions were made for terms such as *witcraft* for 'logic', *forespeech* for 'introduction', *bearder* for 'barber', *flesh-strings* for 'muscles'.[23]

The field of botany presents a good example of the problems of naming and defining; it also and particularly raises the question of specialized knowledge and the specialized linguistic devices necessary to represent it. Before Linnaeus put things in order once and for all, the linguistic disorder faced by the seventeenth-century botanists was as bad, if not worse, than that faced by the population at large. Quite simply, for example, plant varieties were rapidly being discovered and these had no names, or the names were unknown to the discoverers. Or what is perhaps even worse, many known plants had a number of vernacular, local names which varied from country to country and from place to place. Often, there were many synonyms. Plants 'known' through ancient and medieval authors had Greek names or Latin names, or Latinized Greek names, and there was often considerable variation between texts. Quite often the specimens were not known and there was difficulty linking up the name in the text with any known specimens. Vernacular botanists had problems also. Herbalists, for example, found it difficult to write in English. Gerard notes that there are many more Latin botanical terms than the 'vulgar toong' of English could express.[24] When herbalists gave new names to new plants, they alleviated some of the confusion by regularly assigning them Latin names. Latin came to be (or remained) the universal language of botanists. There was, however, no regular system governing the composition of the names. Except in a relatively small number of cases, the botanists had no common language of names with which they could communicate.[25]

This situation accounts for the laborious efforts on the part of many botanical scholars to provide detailed lists of synonyms of plant names to accompany their descriptions. An example from Ray's *Catalogus plantarum circa Cantabrigiam nascentium* makes the point: Ray lists *Anemone nemorum* (taken from Bauhin's *Pinax*) as 'Single white wood crowfoot, Wood anemone or Wind-flower', *Anonis* or *Ononis*, from Johann Schroder's *Medico-chemical pharmacopoeia* (Lyons, 1649), is 'Rest-harrow, Cammock, petty Whinn'.[26] If the synonyms were not given in the plant descriptions, they were given in some kind of lexicon appended to the main body of the text. In addition to these lists, we also find polyglot dictionaries of plant names, e.g. William Turner's *The Names of Herbes in Greek, Latin, English, Dutch and French with the Common Names that Herbaries and Apothecaries Use* (London, 1548).[27]

Plant descriptions, which we looked at in the last chapter, increasingly became pared down along morphological lines, and served as definitions. The major debate in the first half of the century centered around which characters to include in an essential definition of the plant. Rationalized description came to fulfill the function of defining.

Clearly, the need for a rationalized language was as great, if not greater, for the scientists as it was for the linguists. Although the system of plant nomenclature was not finalized and the rules for name formation were not articulated until the time of Linnaeus, creation of some kind of rational system of naming was a necessity long before his time. Botanists through the seventeenth century each had a hand in it. Put in this perspective, the projectors were simply one of many groups engaged in this necessary construction of a specialized language. What they took from the naturalists and botanists was the model of the self-defining language of taxonomy.

ii. Botanical nomenclature and taxonomic names

Unless one wishes to carry a bag full of things as Swift's Laputians did, it is necessary to have words. In the seventeenth century it was understood that the primary function of a name is to refer or to designate.[28] It names a thing or a group of things, and enables one to 'point' to or mark out the thing in question despite the fact that the thing is not present in time and space. The significand of a name can be many things: it can be mental – our ideas, feelings, fantasies; it can be what we have experienced or believe ourselves to have experienced in nature – horses, petunias, clouds, as well as fairies, choirs of angels, the man in the moon (and his lanthorn and dog), gravity, atoms and photosynthesis. In the seventeenth-century theory of desirable designation these possibilities were severely restricted: names were to represent things, but only those things of which we have empirical, visual knowledge. Language is to be created out of what can be seen (or what can be experimentally verified, which is another way of seeing). What can be seen are things and their characteristics and behavior. This is why so much of what is meant by language in the seventeenth century should in fact be called 'nomination'. The representation of the visual is primarily a list of names, a nomenclature.

A name associates a sign with a thing; and included here also are concepts for things. The name acts as a mnemonic device to enable us to recall the significand. But in the act of referring, a name simultaneously recalls not only the thing itself but also what is known about the thing. This field of knowledge necessarily varies from age to age and person to person. The knowledge, for example, that we find associated with plant names in earlier periods – folklore, legend, magical powers, medicinal uses – would be different from the knowledge recalled by the same name in a seventeenth-century botanical treatise. There, names are used to express, to the extent that it is possible, knowledge of the structural features that constitute the form and lead to the essence of the thing. At the same time, these names express knowledge of generic relationships. Thus, through language – a regulated language such as the linguists and botanists were to create – 'there is revealed the entire order of nature'.[29]

Philosophers might claim that natural language also constitutes such an order; poets (and madmen) might argue further that natural language reveals relations far more significant and far richer than any language of taxonomy. But, for many in the seventeenth century, natural language presented a piecemeal and fragmented representation. The time was ripe to create a language of systematic representation, a language of taxonomy.

As we have pointed out, classification is natural to thought. It is not a question of whether we will classify but rather *how* we will classify and to what extent. Human language, be it natural or artificial, will reflect these various kinds and degrees of classifying. It is not unusual, therefore, that, as we see rational taxonomy growing out of folk classifications, we see also systems for rational naming growing out of patterns existing in ordinary everyday language, specifically, out of the practice in ordinary language in which some names tell something about the properties of the thing, e.g. *great white water* lily. We see in botany the similar construction of a nomenclature in which names tell the genus, species and difference of things. These kinds of names, essential names, both express the taxonomic ordering and state the essence of the thing. It was these two criteria – that names express taxonomic order and that names be self-defining – that the projectors adopted for their artificial languages. These two criteria guaranteed that what they named and defined was the (real) essence or nature of things.

We can describe certain dimensions of plant-naming in ordinary language at a particular point in time. One such dimension is showing taxonomic relations. Amongst the generic names some, the most 'primitive', will be single names, and these obviously do not designate taxonomic relations, e.g. *oak, orange, iris, pansy, coriander, lily, pear, plum, apple.*[30] Included here also should be names formed by metaphors where, although there is more than one word, the name is considered to be a single unit, e.g. *snapdragon, bachelor-button, jack-in-the-bush, cat-tail, beggar-weed.* There are also generic names composed of two words; these result from (1) the creation of a generic name on analogy to an already existing name, e.g. *mock orange, water lily, poison oak;* or (2) they come about after higher categories have evolved, e.g. *planetree, blueberry, raspberry, strawberry.*[31] In this latter case a relation with the higher class is expressed but this is not important enough to be a regular, necessary pattern applied to each variety.

Specific names are composed of two parts (not necessarily limited to two words) and express generic relations except in the following important exceptions: (1) in the case of monotypic genera, where the specific name is identical to the generic name and there is no need to append a difference; and (2) in the case of the species which are considered to be the most typical or most important where often there is only one, generic, name; they may later take on epithets like *common, true,* etc.[32] In all other cases, specific names have one part that states the genus and another part that indicates that each species is different from its congeners, e.g. *jack pine, Granny Smith apple, Easter lily, calla lily, tiger lily, live oak, blue spruce, white water lily.*

Varietal names also express relationships: they consist of the specific name which indicates the common group and a descriptive epithet which distinguishes the varietal congeners, e.g. *baby lima beans, butter lima beans.* These are often shortened, e.g. *baby limas, butter limas.*[33]

Besides expressing generic relations, plant names may or may not be transparent, i.e. may or may not tell something about the nature of the plant. *Pear* and *reed* are not at all transparent; *tiger lily* and *strawberry* are partially transparent because the specific epithets, *tiger* and *straw* at least indicate that these are different from other members of the genus. *White water lily, swamp white oak, oriental planetree* and *blue spruce* are fully transparent. (There will always be at least one non-transparent, root element.) There are no doubt many degrees of transparency, and most names will fall between the two extremes.

It is conceivable that generic names as well as specific names could be transparent. *Blueberry* is partially transparent, more so than *planetree*, which is more transparent than *water lily* and *mock orange* because the planetree is a tree while the latter are not lilies or oranges. Linnaeus cites *triopteris* (three wings) as the case of the ideal (transparent) generic name: it appropriately represents a genus which has three petals ('wings') standing around its inner petals.[34]

One thing that complicates this simple scheme is the fact that the meaning of words, and hence the transparency, may get lost (or may be lost on a particular group of speakers). This is particularly true for foreign borrowings but is not limited to them. A further complication arises because most words will change phonetic 'shape' in the course of time, and this too may affect the transparency.

Rose comes from Latin *rosa* which comes from Greek *rhodon* which means 'rose'. Here, a rose is a rose is a rose and its meaning is not transparent in any language. *Orchid* on the other hand ultimately derives from the Greek word *orchis* meaning 'testicle' and the plant is so named because of its similar shape. The name in Greek had some degree of transparency. The same is true for *pansy* from French *pensée*, 'thought', and *gladiolus* from Latin 'little sword'. *Heliotrope*, from the Greek *helios* and *trepein*, the words for 'sun' and 'turn' respectively, is fully transparent.

Showing generic relations and having transparency are two different functions of names and there is no necessary relation between them. Some words neither show relations nor have transparency, e.g. *pear*. *Triopteris* is transparent but does not show relations. *Tiger lily* and *jack pine* show relations but are not particularly transparent. *White water lily*, *blueberry*, *blue spruce* show relations and are fully transparent.

When a (species) name both shows relations and is fully transparent the name is self-defining (where species is always a subordinate class). All names therefore are either (1) simply designatory, or (2) both designatory and self-defining. In the self-defining name, the specific epithet in the definition names the characteristic of the species that distinguishes it from its congeners. At early stages of naming in ordinary language this is almost always the case. Aristotle tightened this up by making it necessary, and it may well be that he derived his rules from practices he saw in natural language. Strictly speaking, in Aristotelian botany, the specific characteristic should indicate something essential to the life of the plant. *Swamp white oak* and *blue spruce* are self-defining names insofar as *swamp* and *blue* refer to characteristics that differentiate the species, and are, conceivably, related to the essence of the plant. *Tiger lily* and *strawberry* are not self-defining since *tiger* and *straw* do not transparently state the nature of the essential differences. In modern botany, self-defining specific names are called essential or diagnostic.[35]

The first part of a diagnostic name is composed of one or two or more words, which furnish the common denomination, i.e. state the genus. The second part is a phrase of varying length that states the characteristic of the species. Theoretically, the function of the second part is to state the essential difference. As we have seen, however, determining the essential character was exactly the subject of botanical research. Varietal names follow the same pattern: one part of the name states the difference; the other part names the common group.

As the study of plants developed into a rational, specialized discipline there was a tendency to create plant names so that the species and varieties were marked by diagnostic essential names – so that these names showed generic relations, were transparent and were self-defining. This continued throughout the seventeenth century and into the eighteenth century when it became institutionalized by Linnaeus. The *binomial nomenclature* for which Linnaeus is famous is not, however, a set of diagnostic essential names; it is a set of designatory names. Linnaeus created first a nomenclature of diagnostic names and then he created another, parallel set of designatory names: these were the true binomials. These designatory names consist of one word which names the genus and a second word which differentiates the species. These specific epithets, called 'trivial names' by Linnaeus are not, nor need they be, descriptive of the specimens; they are very often personal names, place names, etc. Designatory specific names do not convey descriptive information about the plant; rather they simply indicate that the various species belong to a common genus but are separate groups within that genus. They label the species in question but tell nothing about its qualities. The same holds true for most generic names. Although Linnaeus thought it preferable that the generic name be self-defining (e.g. *triopteris*), it was not possible to work this out to any great extent; most names for genera are in fact designatory. The function of a designatory name is to show generic relations and to show the limits of the class, i.e. to show identity and difference between specimens. These are to be contrasted with diagnostic names which, because they are tied to taxonomic tables and the information contained therein, indicate the arrangement and definition.

The following is an example adapted from Linnaeus' *Species plantarum* (1753)[36] which shows both the diagnostic and the designatory, trivial names.

trivial name	*species no.*	*generic name*	*differentia = true specific name*
calamogrostis	5	Arundo (reed)	calycibus unifloris, culmo ramoso (one flower calyces, branch-like stems)
arenaria	6	Arundo	calycibus unifloris, foliis involutis, mucronato-pungentibus (one flower calyces, rolled leaves, sharp stinging points)

The designatory binomials, *Arundo calamogrostis* and *Arundo arenaria*, show that the species belong to the same genus but that they are different. If we want to communicate, the designatory names are sufficient and efficient. Furthermore, they are easy to memorize. If we want the definition of the species, we must refer to the list of diagnostic names. Designatory binomials separate the functions of naming and defining; diagnostic names combine those functions.

From the time of Theophrastus until the time of Linnaeus' nomenclatural reforms, botanists tried to give diagnostic names for plant species.[37] Theophrastus did so by virtue of the demands of Aristotelian philosophy, i.e. by the necessity of giving definitions by genus and difference. Others did so either because they were similarly followers of Aristotle, or because this was the tradition established by Greek and Latin texts, or because diagnostic naming is simply an extension of one kind of naming found in vernacular language.

In the works of Theophrastus we can see the process by which diagnostic names are formed. Below is an example in which descriptive terminology is made into specific epithets:

> The willow also grows by the water and there are many kinds. There is one which is called the black [specific epithet] willow because its bark is black [descriptive term] and red, and that which is called the white [specific epithet] from the white [descriptive term] color of its bark. The black [specific epithet] kind has boughs which are fairer and more serviceable for basket work while those of the white [specific epithet] are more brittle.[38]

Theophrastus took his specific epithets from various features that were thought significant to the plant, e.g. color, habitat, size, shape, use. Below are examples of the patterns he used for diagnostic names:

Raphanos (cabbage)	Raphanos oulophyllos	curly-leaved cabbage
	Raphanos leptophyllos	flat-leaved cabbage
	Raphanos agria	wild cabbage
Kalamos (reed)	Kalamos auletikos	reed used for making pipes
	Kalamos charakias	reed used for stakes
Thridakine (lettuce)	Thridakine leukos	white lettuce
	Thridakine platykauloos	lettuce with a flat stalk
	Thridakine strongylokauloos	lettuce with a round stalk
	Thridakine lakonykon	Laconian lettuce
	Thridakine melainos	black lettuce[39]

In the seventeenth century, plants were still being named in this fashion except that as the number of species increased, the names of the plants grew longer. Bauhin's entry under the genus *Apium* provides a typical example. Listed under the generic name are fifteen specific names, each of which is composed of the name for the genus and a descriptive phrase (specific epithet). It will be noted that the generic name is not the same in all cases; in early botanical works it was the custom to keep the names given by the ancient authors even when revisions in the taxonomy regrouped the plants.

Apium

1. Apium hortense or Petroselinum vulgo
2. Apium or Petroselinum crispum
3. Petroselinum Creticum
4. Apium hortense latifolium
5. Apium montanum folio tenuiore
6. Apium montanum folio ampliore
7. Apium sylv. lacteo succo turgens
8. Apium montanum nigrum
9. Apium peregrinum foliis subrotundis
10. Apium peregrinum cicutae folio
11. Apium palustre or Apium officinarum
12. Apium Macedonicum
13. Hipposelinum Theophr.
14. Smyrnium peregrinum rotondo folio
15. Smyrnium peregrinum rotondo folio oblongo[40]

In Bauhin's entry for *Sison*, we have a monotypic genus and the specific name is simply *Sison*.[41] With *Pastinaca latifolia* and *Libanotis ferulae folio* (as distinguished from *Libanotis*) we have generic names composed of more than one word.[42]

The specific epithets constitute descriptions of the species. The primary descriptors are location, e.g. *montanum*, *Creticum*, *palustre*, *Macedonicum*; or how it grows, e.g. *hortense*, *peregrinum*, *crispum*. A large number includes morphological descriptions, e.g. *latifolium, folio tenuiore, folio ampliore, foliis subrotundis; cicutae folio, rotondo folio, rotondo folio oblongo*. Other properties are mentioned, e.g. color, *nigrum*, and shape, *lacteo succo turgens*.

John Ray's first work, a catalogue of Cambridge plants, displays a similar tendency to use diagnostic names, but as is the case with all the previous herbalists, they are not regularized or systematized. Ray's plants are given in alphabetical order rather than being classified. *Cicuta* as a separate listing is followed by *Cicutaria fatua; Cicutaria palustris; Cicutaria vulgaris*. The genus *Conyza* has six separate entries: *Conyza major; Conyza media; Conyza minor; Conyza coerulea acris; Conyza foliis laciniatis; Conyza palustris*.[43] Thus Ray and Wilkins were admirably suited for collaboration on Wilkins' philosophical language. Both the scientist and the linguist were each seeking a way to establish regular, self-defining, essential diagnostic names for the order of things.

John Ross once prefaced a linguistic study with a tale about a philosopher who was told that the world rests on the back of a turtle. The clever philosopher asked what the turtle rests on and was told that it rests on the back of *another* turtle. To cut an (infinitely) long story short, it was turtles all the way down. What Ross found disconcerting in the story but nevertheless appropriate to a study of language was that there was no beginning (or end if you like). Most of us know the feeling.

Not so however with the seventeenth-century language projectors. As followers of Aristotle, language had a bottom. Written words rested on spoken words or names, names rested on ideas of species, ideas rested on taxonomic structure, taxonomic structure rested on the species of things, the species of things rested on their essential characteristics, the essential characteristics rested on the essences. The beginning was in the solid bottom of the *prima materia* and the unmoved mover, the final turtle with feet planted firmly on the ground. In an unbroken chain of succession (not unrelated to genealogy, which is not unrelated to taxonomy) these parts formed the links in the taxonomic episteme, an episteme that makes it possible to relate a theory of nature on one end with a theory of language on the other end. Conceive of it if you will as a ladder. Its foundation was stable and from that first step the seventeenth-century virtuosi, linguists and scientists alike, started their ascent into the advancement of learning, the development of science and the creation of a real character and philosophical language.

PART II

The development of the universal languages

CHAPTER 4

The Groundwork

i. Words and things: Bacon

In order to do specialized, scientific taxonomy a unique, decontextualized view of nature must be developed. What we can see in the universal language schemes as they evolve towards taxonomy is the same progression of conception, understanding and apprehension that we saw in the study of plants. But it did not start out this way. In the early stages the projectors considered nature as a living discourse and ordinary language as a discourse discursively representing it. The first universal language projects were simply projects for a new written representation of this ordinary, discursive language.

In the initial stages of development, when one spoke of a universal language, e.g. Bacon, Wilkins and a handful of forgotten Frenchmen, by 'language' they meant an ideographic writing system similar to Chinese characters (as they were then understood), Egyptian hieroglyphics, algebraic symbols, shorthand, numbers, ciphers and the like. The written ideographic symbols were called real characters and were thought to be universal insofar as a sign referred to a notion, thought to be universal, and not a (phonetic) word. The point of a universal language based on real characters is that it is a non-alphabetic writing system meant to circumvent the (spoken) words of disparate languages; it represents not sounds but rather units of discourse – morphemes, words, things, notions.

The development of decontextualization and of a universal language begins with Bacon, just as for botany it began with the late herbalists. Bacon was the first to propose the institution of a real character analogous to Chinese characters. He claimed on the one hand that an ideograph was not a representation of a spoken word but rather it was a representation of an object or notion – a kind of stylized picture if you will. But, at the same time, it is also clear that these characters represent words, albeit silent ones. What is at issue here is that words and things/notions are inseparable in the same way that taxonomic names/structures are.

That the very first schemes were writing systems (real characters) does not preclude the possibility however that these writing systems were also implicitly related to other levels of language, something the cleverer projectors came to understand. The relation becomes more apparent if we consider for a moment the nature of alphabetic representation. While it is commonly believed (and is no doubt true in many ways) that alphabetic writing is a system in which individual signs represent individual sounds, it is also useful to be aware that an alphabetic writing system is not purely a phonological transcription, is not 'merely a way of recording language by means of visible marks'.[1] Rather it is 'separate and potentially an independent system of symbolization which could be used without reference to the spoken language'.[2] With widespread literacy and standardized spelling, 'the written languages of Europe, which were once purely phonetic, are now to a certain extent ideographic'.[3] 'In consequence of our common practice of reading, written language has to a great extent become an instrument for the direct expression of meaning, co-ordinate with audible language. The result of this has been that the written language has in part been developed on lines of its own, independent of the development of oral speech.'[4] Literate speakers, those who read, write and spell, accustom themselves to purely written forms of the language and 'by long habit have formed direct associations between certain familiar groups of letters and the meanings which they represent'.[5] Written words are processed as units and become ideographs for words and morphemes.

In the sixteenth and seventeenth centuries, natural languages like English were being subject to the 'normalization of spelling' and their speakers were experiencing vast increases in literacy, i.e. in the number of readers and spellers. (We might note that all this is predicated on the emergence of a standard language.) This means that while those speakers were using an alphabetic system they were also becoming introduced to or accustomed to a partially ideographic kind of written representation. They were experiencing the process by which what is, in theory, a phonological script was being transformed into a quasi-ideographic form of writing. If this is what was happening in the mind of literate man, it is not surprising that he thought of other, more ideographic writing systems, especially since the example of Chinese was fresh. Furthermore, if written language was accustoming speakers to process written language ideographically and morphemically, it should not take long to see the connection between the written form and the words and morphemes of the language.

Thus we start this study with Bacon and with his remarks about a real character, remarks that we see rising from a widespread involvement with written representation brought on in a context of increasing literacy. That real character in turn suggests that languages can be reduced to a set of root or primitive words. This presupposes that there is a common set of notions – like numbers – independent of, but stored in, natural languages which the ideographic signs are to represent. Understanding the analogy between number signs and ideographic word-signs, and armed with the belief that such a common set of things and notions does indeed exist, the language projectors set out to determine those simple notions in taxonomic tables and to represent them in an artificially constructed language.

The sixteenth and seventeenth centuries were heirs to a world view that is essentially Aristotelian and that world view had repercussions in the study of language; it created the way in which people thought about language and in turn provided the model for the way in which they analyzed language. What we see beginning in the sixteenth century and gaining force in the seventeenth century is the Aristotelian–Scholastic idea that language is to be understood as a thing of nature just as a plant is, and is accordingly subject to logical, Aristotelian analysis. More specifically, concepts of language and grammar are brought into line with Aristotelian metatheory.

There are many facets to this linguistic development which have been admirably studied by Salmon and by Padley.[6] Where these theories interest linguists particularly is in relation to morphology, syntax, and to notions of deep and surface structures or universal and specific grammar. For our purposes however what is interesting in this development of sixteenth- to seventeenth-century grammatical theory is the way in which words are understood to be related to things.

In the tradition of Aristotelian–Scholastic grammar, a theory of language was built on an epistemological theory. It was assumed that the things of nature are the same for all men. It was assumed further that the concepts we have of things are faithful representations of reality, that the understanding mirrors things themselves, that there is a 'conformity of things with the understanding'. It follows that not only are the things of nature the same for all men but the concepts of those things are the same for all men. This position has sanction in Aristotle: 'As writing, so also is speech not the same for all races of men. But the mental affections themselves, of which these words are primarily signs, are the same for the whole of mankind.'[7] This repository or set of universal 'mental affections' is the universal component of the universal languages, the universal, mental deep structure of which individual languages are but the less interesting surface manifestations.

The next step in the argument makes an equation between language/words and those concepts; words are considered to be a faithful representation of phenomena and language in its totality in discourse – a reflection of the relations between them. Just as the particular parts of plants were put together by the syntax of the organism, words are put together by the grammar of the language. And just as the botanists argued that the visible, structural characteristics of plants were manifestations of the essences of plants, so words, sentences and discourses are the observable signs of our ideas of things and the judgments we make of them. As Comenius was later to put it, things are the kernel, language is but the husk. What we have is an unbroken chain between essence, thing (species), idea, word/name, writing. It is because of this linking that words are so frequently treated as if they are the things they represent and that real characters could be thought to be interchangeable with things. It was not until scientific theory overthrew Aristotelian physics that this relationship was weakened.

As Padley makes clear, this Aristotelian grammar flourished under the medieval grammarians.[8] It began to return in the sixteenth century, tentatively in the work of Melanchthon and more clearly with Scaliger and Sanctius. In the seventeenth century it came into its own once again with the universal grammarians, Vossius, Caramuel and Campanella, as well as with the universal language projectors.

The universal grammarians are concerned with the analogy between words and things or the nature of words and the nature of things, a concern which shows itself primarily in their discussions of the parts of speech. But they were also and more especially concerned with the way that words and notions are joined together, a concern which leads them more into considerations of word derivation, morphology and syntax (grammar).[9] There is no question but that the universal language projectors were influenced by these theories both generally and specifically and that they were concerned with universal grammar. Bacon discussed it; Dalgarno and particularly Wilkins devoted large sections of their language projects to formulating principles of universal grammar (Wilkins explicitly on the model of Campanella). But those grammars deal with complex notions and their combinations in propositions and sentences. Those complex notions and propositions, however, in turn rest on the identification of the simple, common notions, elements, or things of nature. The identification of those is the point of taxonomy.

Beginning with Bacon, the language projectors argued that ideally a language should have one word for each basic, simple thing or notion, and have no more than that. Not many words for one thing, or one word for many things, or words for things that do not exist, or no words for things that do exist. This is what Aristotle meant by unequivocal words which he took to be the point of departure in logic and in science. Bacon showed that this parallelism had broken down – because of the difference between the understanding of the learned as opposed to that of the vulgar – and that words did not truly, accurately and unequivocally represent the phenomena of nature. He called for rectification of the disparity between words and things as did later projectors who saw one of the functions of a real character and artificial language as putting the anomalies aright (this came to mean associating one word with one essential definition). We could conceive then of real characters and rectified words as constituting parallel and interchangeable media. The form of the representation is insignificant; what is significant rather is the projectors' understanding that there need be created some medium in which the order and systematicity of nature could be figured forth.

Behind Bacon's simple suggestion that a universal language be created from a real character lay a whole episteme – one founded on an Aristotelian notion that the basic elements of nature are essences, that these sort themselves into kinds, that the proper study of nature amounts to an arrangement or classification of these kinds, that these basic essences or units can be ideally represented by a set of radical names and these graphically by a set of ideographic characters, that in the current state of language this ideal did not obtain, that there was no isomorphism between words and things, that this is the result of the advancement of learning which necessitated an overhaul in the old taxonomy of the Great Chain and the creation of a new vocabulary. The evolution of the artificial languages, beginning with Bacon, is the making explicit of the totality of the episteme.

In *The Advancement of Learning* (1606) and its expanded Latin version, *De dignitate et augmentis scientiarum* (1623), Bacon discussed the art of communication and the signs which make it possible. It was here that he introduced the idea of a real character.

The Advancement of Learning might have been alternatively entitled the anatomy of learning and as such it was a typical Renaissance compilation of information related to the encyclopedia tradition and, as we shall see, not unrelated to taxonomy. In the *Advancement* Bacon systematically described, analyzed and classified Renaissance provinces of knowledge, dividing learning into the arts of imagination, memory and reason, a common Ramistic classification. Among the subdivisions of the arts of reason was rhetoric; among its parts Bacon included the Art of Transmission, or communication.

Bacon observed that, taken in its widest sense, communication is not coterminous with speech but that it 'has some other Children besides Words and Letters'. Nor are the formal units of communication limited to the verbal; they are 'whatever can be divided into differences sufficiently numerous to explain the variety of notions (provided those differences be perceptible to the sense)'.[10] Among these are visible symbols or signs.

Bacon called non-verbal signs the 'notes of things'; these have 'signification without the help or intervention of words'. Some of these notes are 'ex congruo, where the note has some congruity with the notion'; others are 'ad placitum, where [the note] is adopted and agreed upon by pleasure'. The class of congruous notes is made up of gestures and hieroglyphics; they 'have always some similitude to the thing symbolized, and are a kind of Emblem'.[11]

Among the notes agreed upon by convention, Bacon cites alphabetic letters as well as the newly discovered Chinese characters. The latter he called 'real characters, not nominal'. As he understood them, these were characters 'which represent neither letters nor words, but things and notions; insomuch that a number of nations whose languages are altogether different, but who agree in the use of such characters...communicate with each other in writing; to an extent that indeed any book written in characters of this kind can be read off by each nation in their own language'. These real characters differ from other signs in that they are representations of notions, concepts and/or silent words (surds). In a language utilizing this type of written character 'there ought to be as many real characters as there are radical words'.[12]

As a kind of visual or written symbolization, real characters struck many in the seventeenth century as ingenious, but it was not merely the idea of real characters that inspired the Baconian language projectors. If written signs stand for words, and words are ultimately the means by which notions of the real world are communicated, a proper system of (written) signs rests on a proper account of that world. Language, as Bacon claimed, should rest on our proper understanding of nature, and that understanding came, not through books or from authority, but through scientific observation. Reform of language, and/or the creation of an ideal, artificial language, had to be based on science. This was the source of the inspiration.

This leads to Bacon's (empirical) scientific method. Bacon insisted that natural philosophy could develop only if men devoted more time and attention to the observation of nature itself; that, in the past, learning had been concerned more with verbal truth than with the reality of nature. This verbal truth encompassed both high and low: it was the truth passed on by virtue of the authority of the ancient writers and the kind of truth passed on in language and in folk tradition. In Bacon's view, ignorance about the nature of things is what produced linguistic inaccuracy and vice versa. The philosophers of the past, especially the Schoolmen, bound to their authoritative texts and confined to their disputations in the Halls, had turned inward into their logical and rational minds for the discovery of truth. But, for Bacon, the mind was radically incapable of making that discovery unless aided by information from nature itself. According to Bacon's reading of history, from these confused and shadowy regions of pure thought, transmitted through words and untested by experience, there had come into being, and subsequently into popular acceptance, the Idols or false notions of the human mind.

What concerns us here are the Idols of the Market Place, the false notions of the vulgar that come to us through discourse, 'idols which have crept into the understanding through the alliances of words and names'. The reason for these false notions is that words are created by ordinary men and 'follow those lines of division which are most obvious to the vulgar understanding'. These need not, and most usually do not, resemble the deeper understanding of the scientist or philosopher; words are not isomorphic with the nature of things. Once words come into existence, they plainly force and overrule the understanding for the learned and the vulgar alike; subsequently the Idols of discourse prevent the advancement of learning: 'whenever an understanding of greater acuteness or a more diligent observation would alter those lines to suit the true divisions of nature, words stand in the way and resist the change'.[13]

Bacon provides an analysis of the various kinds of anomaly between the things of nature and the words men use, an analysis that was to be oft repeated in the seventeenth century. Some things go undiscovered and are 'left unnamed through lack of observation'. Conversely, there are 'names which result from fantastic suppositions [to] which nothing in reality corresponds'. Among hypostatized words of this sort Bacon cites 'Fortune, the Prime Mover, Planetary Orbits, Elements of Fire and like fictions which owe their origin to false and idle theories'. The most dangerous case, however, is to be found in 'names of things which exist, but yet confused and ill-defined, and hastily and irregularly derived from realities'.[14] The word *humid* serves as an example. According to Bacon it is:

> nothing else than a mark loosely and confusedly applied to denote a variety of actions which will not bear to be reduced to any constant meaning for it both signifies that which easily spreads itself around any other body; and that which in itself is indeterminate and cannot solidise; and that which readily yields in every direction; and

that which easily divides and scatters itself; and that which easily unites and collects itself; and that which readily flows and is put in motion; and that which readily clings to another body and wets it; and that which is easily reduced to a liquid, or being solid easily melts. Accordingly, when you come to apply the word, if you take it in one sense, flame is humid; if in another, air is not humid; if in another, fine dust is humid; if another, glass is humid. So that it is easy to see that the notion is taken by abstraction only from water and common and ordinary liquids, without any due verification.[15]

While Bacon criticizes words here, it is clear in his later works that linguistic problems merely reflect conceptual problems; indeed, we may call them taxonomic problems caused by increasing knowledge. What he criticizes therefore is not only the unreliability of words but of the concepts which words stand for: 'There is no soundness in our notions whether logical or physical. Substance, Quality, Action, Passion, Essence itself, are not sound notions; much less are Heavy, Light, Dense, Rare, Moist, Dry, Generation, Corruption, Attraction, Repulsion, Element, Matter, Form and the like; but all are fantastical and ill defined.'[16]

Our notions of substance are the least problematical but those of action and especially of quality are more difficult:

Our notions of less general species, as Man, Dog, Dove and the immediate perception of the senses, as Hot, Cold, Black, White, do not materially mislead us; yet even these are sometimes confused by the flux and alteration of matter and the mixing of one with another.[17]

One of the least faulty kinds is that of names of substances, especially of the lowest species and well-deduced (for the notion of *chalk* and of *mud* is good, of *earth* bad); a more faulty kind is that of actions, as *to generate, to corrupt, to alter;* the most faulty is of qualities (except as are the immediate objects of sense) as *heavy, light, rare, dense and the like.*[18]

It is at this point in Bacon's philosophy that language and science come together; it is also the point at which present inadequacy and future reform are mapped out. If language simply represents 'the popular notions of things' and if notions have only 'confused relation to the material world' it follows that 'the whole logical edifice crumbles'. Once the crux of the matter had been identified, 'this linguistic view had two marked effects upon later scientists. First it determined the most consistent charge which they brought against the philosophy of the past, namely that it was merely verbal, with no counterpart in reality; and second, it influenced them to consider a reformation in language... as essential as one in science...'[19] There can be no accurate communication including certain, scientific communication unless words transparently represent the true nature of things; and if language or words are to be 'true' they must mirror our observations of the properties of things and of the ways in which things behave. Language and the model of nature must be isomorphic. What we shall see is that this model is deemed to be taxonomic and while this 'reformation of language' came to include mathematics, for a taxonomic model it is words that are important.

This requirement of isomorphism between language and nature leads directly to Bacon's chief contribution to scientific reform, his development of a scientific method which begins with data collection and empirical observation. This observation however is not the end of Bacon's program of scientific reform, nor should observation be understood to be merely sense perception. Bacon had a profound distrust of the ability of the senses to determine anything significant without the aid of the understanding and hence he insisted on the need for experimentation.[20] But while his method begins with modern ideas of observation and experimentation, it nevertheless ends with the classical procedures of classification and definition.

Bacon's scientific method starts with the collection and enumeration of 'instances' and ends with an accurate definition of the nature of those instances. This is not only the paradigm of Aristotelian science but also the paradigm of the science of a literate culture, a culture in which large amounts of data could be collected, preserved and transmitted through the medium of writing/print.

Enumeration was to be followed by classification, the critical operation in the discovery and the definition of the essences of things. Natures could be properly defined only when instances were properly classified into 'the true divisions' of nature. As Bacon insisted, the determination of those divisions was an empirical and practical matter. Rather than speculative, intellectual analysis, it required observing the properties of things and matching those properties with the classes. Bacon's linguistic theory followed from these methodological requirements: words (names) would be accurate insofar as they mirrored the empirically determined divisions of nature and insofar as their limits matched up with those found in nature.

That observation and classification are the cornerstones of Baconian science becomes clear in the most methodological of his works, the *Novum organum*. In order to justify his new method (organon) Bacon's first task is to show the inadequacy of the old, Aristotelian method of scientific demonstration. He does this by rejecting the syllogism as an instrument for the acquisition of certain knowledge, arguing that the syllogism rests on the shaky foundation of false judgments about the nature of things: 'The syllogism consists of propositions, propositions consist of words, words are symbols of notions.

Therefore if the notions themselves (which is the root of the matter) are confused and over-hastily abstracted from the facts, there can be no firmness in the superstructure.'[21]

The chief problems arise with the premises which according to Bacon amount to little more than *a priori* definitions where the predicates of the propositions follow linguistically from the subjects. For Bacon, a syllogism must not simply match up definitions between terms; the categorization and classification of the predicates (which are the bases for propositions) must accurately represent acts of nature. If they do not, the proposition or syllogism only makes a statement about words. Since the real conditions of things and of nature are not by any means obvious, propositions formed in the absence of experimental or empirical data deal only with the linguistic or the superficial. Unless propositions are based on observation and are verified, the syllogism becomes an empty form which churns out verbal proofs. Aristotle and his followers, Bacon claims, were 'always more solicitous to...affirm something positive in words than about the inner truth of things'.[22] They fashion 'the world out of categories'.

Mathematically inclined scientists later rejected taxonomy as a proper way of explaining nature. What is important to note here is that Bacon did not. What he criticized in Aristotle was not the model of taxonomy itself but rather the end product of Aristotelian classification where the categories or divisions are set forth 'as if they were complete in all their parts and finished...'[23] These are misleading in that they 'seem to embrace and comprise everything which can belong to the subject. And although these divisions are ill-filled out and are but as empty cases, still to the common mind they present the form and plan of perfect science.'[24] For Bacon, the facts of nature were not yet sufficiently known to construct an accurate taxonomy. It is important to note, therefore, that Bacon does not object to the notion of 'cases' but to 'empty cases', nor to divisions but to divisions that are 'ill-filled out'.

Bacon's new empirical method would remedy the deficiencies of the Aristotelian method by requiring observation and experimentation before the classification of phenomena and the framing of definitions. As in the case of Aristotelian science, however, classification and, even more so, definition are the ends of his scientific activity.

While Bacon is identified with the new science and the scientific revolution, there is evidence that his acceptance of classification and definition as the end and means of science can be traced to an implicit acceptance of Aristotelian essentialism. In Baconian physics, natural concrete bodies are considered to be a 'troop or collection' of abstract simple natures. These elemental natures are 'as the alphabet in nature or the colours on a painter's palette that combine and produce the variety of faces and figures'.[25] As in Aristotle, a concrete substance can be defined by the enumeration of these simple natures. Gold, for example, is 'yellow in color, heavy up to a certain weight; malleable or ductile to a certain degree of extension; is not volatile...and so on...'[26]

According to Bacon, scientific investigation should be directed to the discovery and definition of these more abstract and simple natures or to discovery of what he called Form. The starting point of that investigation was the collection and examination of data in a series of tables: one kind to determine 'essences and presences', another to record 'absences and deviations', and a third to ascertain 'comparisons and degrees'.

The tables of essence or presence examine similar natures that permanently occur in the widest variety of instances which are not there by coincidence. The example Bacon uses is heat, a simple nature classified as a species of the more general nature or Form, motion. Among the many instances in which it is found are the rays of the sun, fiery meteors, burning thunderbolts, bodies under the influence of friction, organic matter fermenting, quicklime sprinkled with water.[27] Since the Form or nature (e.g. motion) should be absent when the given nature (e.g. heat) is absent, we can check our identification of the common nature by presentation of a negative instance for each affirmative one. Thus, while the rays of the sun have heat, those of the moon and stars do not; the same may be said for burning thunderbolts as opposed to sheet lightning; boiling liquids compared with liquids in a natural state, and so forth.[28] Finally, since 'no nature can be taken as the true form, unless it always decreases when the nature in question decreases, and in like manner always increase when the nature in question increases' we must set up the tables of degrees and comparisons.[29] In the case of heat, we proceed to examine a variety of substances.

In the end, we should be able 'to find such a nature [e.g. heat] as is always present or absent with the given nature and always increases or decreases with it; and which is...a particular case of a more general nature [e.g. motion]'.[30]

In Bacon's schema, the end of scientific inquiry is the discovery of the 'Form' of the simple natures, i.e. 'the true specific difference, or nature-engendering nature or source of emanation...'[31] This notion of form is a complex one and is what links Bacon with Aristotelian essentialism.[32] The problem is that, on the one hand, Bacon seems to regard the Form of simple natures as definitions of the essences of natural phenomena but, on the other, he seems to regard Form as the explanatory rules for the production of such phenomena.[33] The one point of view marks Bacon as an Aristotelian essentialist; the other as a new mechanist.

By way of example, we see Bacon using typical language when he says 'when I speak of Forms I mean nothing more than those laws and determinations of absolute actuality, which govern and constitute any simple nature...'[34] But he also explicitly rejects the Aristotelian idea that the Form is the cause of the simple nature. He would not, he says, be understood to mean 'that heat generates motion or that motion generates heat (though both are true in certain cases), but that Heat itself, its essence and quiddity, is Motion and nothing else; limited however by the specific differences'.[35] Crombie claims that 'Bacon may be concealing in this deceptively scholastic language concepts far removed from the "substantial forms" and real qualities in the sense of scholastic "natures".'[36] But it is also clear, as Carré points out, that Bacon 'constantly reverts to scholastic modes of interpretation, clinging to qualitative theories of being'.[37] Part of the problem stems from Bacon's inadequate knowledge of mathematics and the consequences of applying mathematics to a theory of nature. Thus, as Collingwood indicates, Bacon may have repudiated Aristotelianism in theory yet in practice he constantly substituted 'the classification of qualitative differences for their explanation in quantitative terms'.[38] Without the new language of mathematics Bacon really had no choice; the language of words leads very simply to essences and to taxonomy.

While Bacon may have wished to develop a clearly non-essentialist theory of science, he failed to do so. What he is left with in the end are Aristotelian statements of essence, i.e. definitions by genus and difference. When it comes to the final summation, we see that the 'form or true definition' of heat, for example, amounts to an identification of the genus and a laying out of the differences, each of which is a modification of the one preceding it: 'Heat is a motion expansive, restrained, and acting in its strife upon the smaller particles of bodies. But the expansion is thus modified: while it expands all ways, it has at the same time an inclination upwards. And the struggle in the particles is modified also; it is not sluggish, but hurried and with violence.'[39] This is the classic form of an Aristotelian essentialist definition.

Bacon's scientific and linguistic ideas come together in classification and definition. In the final analysis, what is required for Baconian science is the discovery and definition of essences, the abstract, simple natures of which all else is constituted. Once these basic concepts are classified and defined, we have the basis for a reformed language capable of handling science. Words should be ordered to stand as signs isomorphic to these elements of nature. That those elements are organized in a taxonomic structure means that the reformed language becomes a taxonomic nomenclature. In this state, there would be no disparity between language and reality, between words and things. Indeed, once words faithfully transcribed or reflected notions, the combination of words would amount to the combination of simple notions. Thus it was that, among the early language projectors, who were inveterate followers of Bacon, the establishment of a real character and the redefinition of words, which presupposed a scientific taxonomy, went hand in hand. Classification and definition were Bacon's way; so too would they become the way for universal and philosophical languages.

ii. Taxonomic grammars: Comenius

The discipline of botany progressed from a state in which plants were treated as lexical items in the discourse of life to a state in which plants were decontextualized and treated as objects. The study of plants in the first instance was based on language; in the second case it was based on observation and analysis of the thing itself. In the development of the artificial languages we see a similar shift – a shift from the idea that the new language or character represents (ordinary) language to the idea that the new language represents things. Chronologically, Descartes was the first to make such a proposal when he argued that an artificial language should be constructed on the true divisions of nature, and we will come to this in due course. It is in the work of Comenius, however, that we see this shift in progress. What Comenius brought out was the taxonomic motivation of language and by so doing effected the transition between Baconian theory and the development of the universal languages.

Although Comenius was primarily a linguist, he was also a Renaissance man and as such not yet affected by the specialization of knowledge that later came to divide scientists from amateurs. As an encyclopedist Comenius could therefore concern himself not only with language but with all provinces of knowledge including natural history. His entire pedagogical reform was aimed at teaching students about things rather than about words, as was the current practice in grammar schools. He argued that teaching language, i.e. the names of things, should proceed only after a thorough grounding in the nature of things and their relations. Thus, in his Latin textbooks, things were arranged in very primitive categories and provided with definitions; these categories and definitions were the common ones found in logic and folk taxonomy. While he did not go very far in analysis, in his philosophical work Comenius proposed that detailed studies of nature be undertaken: Pansophia and Panhistoria, complete and specialized taxonomies. He then developed the notion of a universal language which was to be based on that foundation.

As a pedagogue, Comenius began his long career by attacking the problem of Latin language teaching in the schools. In the course of trying to make improvements in curricula, he wrote a set of Latin grammars, and through these grammars devised a method for bringing the study of nature into the classroom. His Baconian concern with understanding the things and operations of nature first hand brought him eventually to write his own taxonomic natural history. Although Comenius was neither a scientist nor a philosopher (perhaps because he was not), his scientific and linguistic notions are based on an interpretation of nature that is explicitly taxonomical and essentialist.

Comenius was not a self-taught Baconian reformer. In his early years, as a student at the Herborn Academy (c. 1611) he studied with Europe's foremost encyclopedists, John Henry Alsted and Joachim Bisterfeld.[40] The prevailing philosophy at Herborn was eclectic, combining the mysticisms of Lull, the Cabala and Bruno with the utopian ideas of Campanella, the logic of Ramus and the scientific notions of Bacon.

The goal of Alsted and other members of the Herborn school was to discover and elaborate the fundamental unity of knowledge behind these diverse methods and theories. The encyclopedia was to be the form for this 'methodological systematization of all things which ought to be learned by men in this life'.[41] Alsted wrote such an encyclopedia, organized by virtue of disciplines or fields of knowledge, in accord with the principles of Ramist logical classification. He is therefore placed squarely in those movements that reflect the information explosion of the Renaissance and the development of a literate culture.

What is significant about Alsted's encyclopedism is that it is a systematic analysis, an orderly arrangement of the totality of knowledge. Although it is portioned out differently, it is not so far removed from a taxonomy based on Aristotelian categories nor unlike Bacon's organization of *The Advancement of Learning* and his 'histories' of various natural phenomena. What Comenius was educated into, therefore, was the idea that nature (or our knowledge of it) constitutes an orderly, unified arrangement of parts. For Comenius it was the business of education and language to present an all-inclusive account of this ordered system.

Comenius began his career as an educational reformer firmly grounded in a Baconian distrust of verbal learning. In his view, it was the disgrace of the age that 'the intellect was scarcely ever nourished by the actual facts but was filled with the husks of words, with a windy and parrot-like loquacity, and with the chaff of opinions'.[42] Education was perpetuating ignorance since it disregarded the fact that 'things are essential, words only accidental; things the body, words but the garment; things are the kernel, words the shell and the husk'. One must not be studied (as was the case) to the exclusion of the other, Comenius argued. 'Both must be presented to the intellect at the same time.'[43]

What Comenius found in classrooms as he travelled across Europe was that education at that time concerned itself not with the study of nature (and its essences) but almost exclusively with the mastery of Latin and ancient literature. (Furthermore, the manner of teaching Latin was irrational and confusing.) The trouble with this emphasis on language is that, when men study words rather than nature, and when their understanding of things is based on words rather than on nature, problems arise as Bacon had foretold: men communicate 'not as from mind to mind the things or senses of things but exchange between themselves words not understood, or little and ill-understood'. While Comenius grants that the problem is to some extent attributable to problems inherent in the nature of language, e.g. to the infinite homonymities between words and things, he claims that by and large the responsibility lies in the way men learn and then use words. The problems arise 'by reason of the almost constant ignorance of [things'] configurations (that is, the inner constitution of each thing)'. For example, when men name God or Angels 'they know not what that, which they name, is in essence, and how it cometh to be'.[44] In his reforms, therefore, Comenius was determined to start with things themselves. He 'thought it not enough to teach what is called white, blacke, hot, cold, a plant, a man, an Angell, Heaven, God, Christ, etc. There is more need of teaching what those things are that the understanding may be rather acquainted with the essence of things than with their names.'[45]

Comenius devoted the greater part of his life to a total reform of language teaching. Assuming that languages are uniform (in some undefined way), like Bacon he devised a 'new method' ('a more natural method') suited to the operations of nature. Knowledge of things was to come before names, knowledge of words before grammatical rules. The method would systematically approach language to extract: '(1) a full nomenclature, that is, a word for each thing; (2) an exact terminology, that is, full agreement on the meaning of words and (3) a full set of laws for regulating language, that is, dependable rules of syntax'.[46]

As had Bacon, Comenius came to see that reform of language (and language teaching) is inseparable from, is indeed contingent upon, a systematic description of nature. He envisioned this Pansophia as the weaving together of a 'single and comprehensive scheme of human Omni-Science (that is, all things under Heaven which it is granted us to know, to say, or to do)'.[47] In this scheme, he saw the created world as a chain of things, a fabric woven in universal harmony, a harmony which was an expression of the benevolence and perfection of the Divine Creator.[48] Hence Pansophia must concern itself not with the surfaces of things but with the essences of them and with their place in God's universal frame which 'displays all in the very order of things, that each may stand in its own place, as it follows from the premises and begets the consequents from itself'.[49]

In his initial attempts to reform language teaching Comenius devised text books which taught the fundamentals of Latin but their guiding principle (and their uniqueness) was that the nature of things was learned as the words themselves were learned. The Latin textbooks were conceived as a kind of taxonomic encyclopedia. To bring order to the world of things (or to reveal the order) words were grouped in categories arranged to indicate generic relations. Definitions or descriptions of words within the classes were to show the unique properties or inner configurations of the class of things and notions in question.

Taken all together, Comenius' textbooks reveal a set of linguistic attitudes characteristic of all the language projectors. There is first of all the belief that homonyms, synonyms and compound words are excrescences on the 'real' language. If these are cut from the lexicon, what is left is a basic vocabulary or a set of root words: 'basic' in the sense that they are the ultimate reduction of the world of things to the essentials; 'root' in the sense that from them all other words, and then sentences, are compounded. The core vocabulary or notions represent objects or sets of objects, their properties, and the basic taxonomic relations between them. Theoretically at least this is no more than a scheme of Aristotelian substances and accidents arranged by genus, species and difference.

Comenius' first book, the *Janua linguarum reserata* (1631) was immensely popular. It was translated into twelve languages, including English in 1632.[50] Following principles of gradation and uniformity, the *Janua* presents 8000 common Latin words arranged in 1000 sentences which grow in complexity. The scientific or encyclopedic orientation of the work is revealed in the division of these sentences into 100 sections, each of which deals with a class of concepts or things arranged in an implicit order. The classes are listed sequentially beginning with Creation and continuing with elements, firmament, fire, meteors, waters, earth, stones, metals, trees, fruits, herbs, living creatures, birds, water creatures, and so on. These are groups that naturally occur in folk culture, and in its mythic expression in the Great Chain of Being. They are also the traditional categories of Aristotelian logic books. The groups express a series of taxonomic relationships which provide simple guidelines for definitions.

For Comenius, differentiation between classes of things is the first step in acquiring knowledge of things: 'When by the help of my Janua Linguarum youth hath learnt to distinguish things from the outside, it should hence be accustomed to explore that which lieth in things, and to comprehend what each thing is in its essence.'[51]

Since the original *Janua* was too difficult, Comenius wrote the *Janua linguarum reserata vestibulum* (1633).[52] This contained 1000 basic Latin words arranged in simple sentences which described the properties of those words. As it was first issued, the vocabulary was divided into a few very large classes or genera, e.g. the accidents of things, the actions and passions of things. These were then subdivided into subgenera as, for example, the actions and passions, which break down into actions of heavenly creatures, of elements, of plants, of living creatures, of man; actions of man is further subdivided into actions of the mind and actions of workmen.

Comenius later added the *Orbis sensualium pictus* (1658) as the most elementary of his texts. It consists of drawings of 'all visible things (to which also things invisible are reduced after their fashion) of the whole world', accompanied by wordlists which give the names and definitions of the objects in the pictures.[53] The pictures are grouped 'in that very order of things in which they are described in the *Janua Latinae linguae*...'[54] Although it looks like a simple little book, it was considered to be revolutionary at the time and many would agree with the opinion of Charles Hoole, the English translator: 'he hath...descended to the very Bottom of what is to be taught, and proceeded (as Nature itself doth) in an orderly way; first to exercise the Senses well, by representing their objects to them, and then to fasten upon the Intellect by impressing the first notions of things upon it, and linking them one to another by a rational discourse'.[55] What is interesting here is the unquestioned assumption that the most basic or fundamental notions are visual ones.

Aristotelian classifications are evident not only in Comenius' language texts but also in the epitome of science which he wrote, *Physicae ad lumen divinum reformatae synopsis* (1633). The topics Comenius proposed to treat there were more or less the same as the classes of words he developed in the Latin grammars: the motions, qualities and mutations of things; the elements, vapors, concrete substances; plants, living creatures, man and angels. And we learn, as we had in the Latin grammars that 'an elment is the first and greatest body in the world of a simple nature'. Further, that 'there are foure Elements, Skie, Air, Water, Earth...The skie is the most pure part of the matter of the world, spread over the highest spaces of the world...Air is a part of the matter indifferently pure, spread over the lower spaces of the world' and so forth.[56] These are the simpler examples and his text goes on for a couple of hundred pages dividing and defining. With this the linguistic and philosophical schemes of Comenius come together: basic vocabularies and sets of simple sentences amounting to definitions become an encyclopedia of (scientific) knowledge.

As the work of Comenius progressed from the practical to the theoretical, his schemes became more grandiose. In conjunction with the scientific nature of his proposed educational reform, he began to compose *Janua rerum reserata*, a handbook of universal knowledge and an extension of his classificatory efforts in the Latin grammars.[57]

It was, however, the *Via lucis* that most fully stated Comenius' Pansophic dreams. No longer would he confine himself to grammar schools but instead would draw plans for universal schools and universal colleges of wise and learned men, along the lines of Bacon's Solomon's House. The universal school is no longer a building but 'no less than our present life'. The colleges become 'the company of Angels and Holy men united in pious witness to the glory of the Creator'.[58] These schools rest on the foundation of universal books, composed by learned specialists presenting 'true and perfectly arranged summaries of all things that can be known'.[59] They are arranged under the headings of Pansophia, Panhistoria and Pandogmatica. The most general, Pansophia, treats 'the whole fundamental condition of things as they are in their ideas (showing how all things proceed backwards and forwards from a single principle along the lines of a single order to one fixed form of truth)'. Panhistoria then follows to unfold 'all the variety of particular things, that is all the particular actions, accidents and issues of things'. Pandogmatica is a critical work which will review all past theories about things to determine 'whether they are true or mistaken'.[60]

Comenius' plans for Pansophic books have an obvious connection with his linguistic schemes, both in the language textbooks and in the universal language which he was shortly to propose. The ultimate book was to gather up 'an inductive history in which all things, which have ever been exactly observed and proved beyond the possibility of mistake to be true, are faithfully collected and set before our eyes; so that by an adequate examination of each one and by comparison of one with another the Universal Laws themselves of Nature may be brought within our knowledge'.[61] He does not indicate how these laws are to be derived, but it is evident that the first step is to discover what a thing is and what it is not. That is also the first step in constructing a taxonomy.

CHAPTER 5

Proposals and schemes for a universal language

i. *The scientific background of the language projectors*

The universal language movement was situated squarely in the scientific world of the seventeenth century. It was in fact seen as a kind or adjunct of science or as a kind of (taxonomic) natural philosophy. This conjunction of interests – linguistic and scientific – is one of the interesting features of the universal languages. It is not my intent to give a definitive social history of the universal language movement, but, before we begin to describe the particulars of that movement, it is worth setting out a brief schema of the scientific interests and associations of the projectors. The fact is that the artificial language projectors were, with minor exceptions, directly connected with the development of science and were in varying degrees either professional scientists or closely related to them. Nor did their work occur in a social vacuum. They were all 'related'; they all knew each other or knew of each other's work, and information about their projects was transmitted freely. What is important to note is that the universal language movement was associated with a group of men who believed that they were making or helping to make advances in scientific learning.

Having said this however it is also necessary to point out that the projectors were not all alike in their involvement with or their understanding of science. Boyle, Ray and Newton are clearly not in the same class as Lodowyck, Dalgarno and Wilkins, who in turn distinguish themselves from Hartlib and Comenius. What we see in the scientific orientation and activities of the language projectors is a recapitulation of the development of scientific theory and specialization in the seventeenth century. At the one end of the century we have Comenius, for whom natural history is part of a Renaissance, encyclopedic concern with knowledge; at the other end we have Newton, so mathematical and so specialized as to be in his time incomprehensible to most; in between we have the members of the newly founded Royal Society dedicated to experimental science.

The language schemes similarly develop along this scientific continuum. They start out as real characters; their purpose is utilitarian, i.e. to provide a simple, universal form of writing. In the second stage of development there are schemes that set up a basic vocabulary of root words. Along with these comes partial analysis of words representing things, which is the start of analysis of things themselves. With this analysis, language begins to become a taxonomic science. This occurs at the same time that natural history begins to develop as the more specialized natural philosophy in the hands of the Royal Society practitioners.

Although the earliest contributions to the universal languages came from France, we know little of their circumstances and contexts. When the universal language movement spread to England however – and it was there that it developed most fully – it was distributed between two clearly defined 'scientific' groups. One was headed by Samuel Hartlib and was the one that included Comenius. The other was associated with Wilkins and eventually with the Royal Society. In a different context, Trevor-Roper has supplied convenient labels for the scientific attitudes of the two: Hartlib's group was interested in Bacon's philosophy of reform or 'vulgar Baconianism'.[1] The other group was interested in more theoretical aspects, or 'pure Baconianism'. Others have designated the two groups as amateurs and scientists, respectively.

What distinguishes the two is the degree to which they were willing to join scientific activity with social involvement. The vulgar Baconians were interested in science primarily as a means of bringing about technological and material development to serve social and/or religious reform. 'At every step, scientific involvement was governed by the criterion of the "public good".'[2] And while Hartlib and Comenius were not themselves Puritans, their reformatory programs and inclinations drew them into close association with Puritan sectaries during the more radical and utopian phases of the revolution.

Wilkins' group moved in the opposite direction – away from reform and away from social and religious controversy. Their kind of science was developed as an intellectual activity with the primary end of contributing to the development of a body of knowledge (this is why it was experimental from the beginning). The attraction of this kind of 'pure' science was precisely that it was not bound up with 'the passions and madness of that dismal Age' as Sprat put it. Members of Wilkins' group, the pure Baconians, were or became professional scientists, although to be sure at times the distance that separated some of them from the vulgar Baconians was more social than intellectual. It is in Wilkins' group, with its orientation to collection of data, experimentation and demonstration of theory, that we find Wallis, Boyle, Hooke, Ray, Newton and other scientific luminaries of the seventeenth century.

While Hartlib is only a sponsor rather than a creator in the universal language movement, he is the perfect image of vulgar Baconianism and as such of a kind of scientific milieu out of which the early universal languages emerged. He was born in Germany and like Comenius studied with the Herborn encyclopedists, Alsted and Bisterfeld, before emigrating to England.[3] One of Hartlib's lifelong interests was educational reform along Baconian lines and he figures prominently in various schemes to revise curricula and teaching methods, particularly methods of teaching language.[4] It was this interest in pedagogical reform that led to his close friendship with Comenius and to his interest in universal language. He followed developments in science with interest, but from the late 1640s it is clear that Hartlib and his colleagues were more interested in gadgets than in experimental science – gadgets that were not only useful but potentially profitable as well. In this respect, Hartlib was in many ways the crank element of the new science which Swift satirized. Among the projects he sponsored were, for example, Petty's instrument for double writing, Dymcock's perpetual motion machine, as well as other schemes involving orchards, cidermaking, beekeeping, silkworms and clocks.[5]

Hartlib's group is also distinguished by its connection with the more esoteric and hermetic doctrines of the day, an interest which has been shown to have flourished among the more radical elements of Puritanism, which in the 1650s Hartlib came to be affiliated with. Hartlib's continental connections placed him in league with hermetical utopianism, the arcana of the Rosicrucians, the mysticism of Boehme, the secrets of cabalism and the alchemical theories of the Paracelsians. It has been argued that both Hartlib and Comenius were in fact Rosicrucians.[6] His scientific interests, as one historian has put it, amounted to a pursuit of 'secrets' all over Europe. His association with this tradition of 'natural magic' shows his fundamental differences from the more professional scientists.

Two important associates of Hartlib were William Petty and Robert Boyle. Petty joined Hartlib's group in the late 1640s and was involved with a number of medical reformers gathered around Hartlib. He also shared Hartlib's enthusiasm for educational and linguistic reform. In 1649 he moved to Oxford with Wilkins where he was known to be working on a universal language.[7]

Boyle joined Hartlib's group in the early 1640s partly because of their shared interest in the reunion of Protestant sects and partly because of his interests in utilitarian science as a means of social reform. He was a member of what he called the 'Invisible College', a group associated with Hartlib and devoted to utopian technological reform.[8] He shared Hartlib's interest in the Paracelsian–Helmontian chemistry Hartlib had imported from Germany.[9] Boyle's conversion to mechanistic philosophy took place around 1653 and in 1654 he also joined Wilkins' circle in Oxford.

The second group, the pure Baconians, were men who ascended the wheel of fortune as the century wore on and who became what can clearly be called the scientific establishment. They gathered themselves around the figure of John Wilkins and were incorporated into the Royal Society, of which Wilkins was the unofficial founder and first Secretary.

Wilkins' scientific interests go back to his student days and his tutorship at Oxford.[10] His earliest publication, *The Discovery of a World in the Moone* (1638), was one of the first promulgations of the new science in England. After leaving Oxford to pursue a clerical career, Wilkins was placed in the service first of Lord Saye and Sele and then Lord Berkeley.[11] Both were Puritan sympathizers which might indicate Wilkins' political or reformatory bent at this time. It was here that he published *Mercury* (1641) which contained one of the earliest universal characters.

Those who make an equation between Puritanism, reform and a spirit of scientific utilitarianism would associate Wilkins with the Hartlib group of the 1640s. We know that by the middle of the decade Wilkins was corresponding with Hartlib, a correspondence which continued into the 1650s until Wilkins dropped his acquaintance with Hartlib. In 1644 Wilkins was employed as Chaplain to Charles Louis, the Elector of the Palatine, then residing at Whitehall. This would perhaps have brought him into contact with Theodore Haak, a friend of Hartlib and a member of the Elector's court. A German refugee like Hartlib, Haak studied at Oxford and Cambridge from 1625 to 1626 and again at Oxford between 1628 and 1631. It is possible that he met Hartlib in 1626 although there is no correspondence between them until 1635. In the 1630s Haak was in and out of England, active in the collection of money for Palatine refugees, a cause that Hartlib was involved in also. Both were part of the circle around Charles Louis, and by 1644 Haak was acting as his official agent. In 1641 Haak formed part of the delegation (headed by Hartlib) which met Comenius in London.

If Wilkins knew Haak from his work at Whitehall, he could also have known Hartlib. It may also be that Wilkins and Haak already knew each other from Oxford (c. 1630), where both had an extracurricular interest in mathematics. It is also possible that Wilkins could have known Haak and possibly Hartlib through his patron Lord Saye and Sele who shared their interest in Protestant refugees. Still another connection could be through Wilkins' second patron, Lord Berkeley, who like the Elector was interested in mathematical matters. There are therefore any number of possibilities that Wilkins had early acquaintance with Hartlib via Haak.[12]

There are others who claim however that Wilkins' more serious scientific intentions preclude a close association with Hartlib, and indeed there seem to be apparent contradictions in Wilkins' career.[13] What is interesting is that the changes between Wilkins' early and later scientific inclinations follow a pattern we notice with many of the professional scientists. Yates argues that Wilkins' *Mathematicall Magick* (1648) shows influences of hermetical figures like John Dee and Robert Fludd, as well as familiarity with Rosicrucian manifestos.[14] His interest in 'artificial curiosities', gadgets and inventions shows continuity with the utilitarian Baconianism of Hartlib's group, interests which for a time were shared by Boyle and Petty.[15] Evelyn describes an occasion when Wilkins showed him 'transparent apiaries which he had built like castles and palaces...These were adorned with a variety of dials, little statues, vanes, &c...He had also contrived a hollow statue, which gave a voice and uttered words by a long concealed pipe that went to its mouth, whilst one speaks through it at a good distance.'[16]

If this associates Wilkins with the more utilitarian vulgar Baconians, there is also evidence that shows him to be a more theoretical scientist. Aarsleff has argued that, right from the start of his career, Wilkins was indebted to Mersenne both for serious scientific theory and for knowledge of Descartes' and Mersenne's notions about a philosophical language.[17] Wilkins was also a member of Theodore Haak's 1645 scientific group, a group founded in order to conduct experimental science and a group to which Hartlib for example did not belong.[18]

By the end of the decade, Wilkins moved to Oxford, taking a number of the members of the 1645 group with him, e.g. Wallis and Ward. Once at Oxford he was responsible for gathering together members of the Oxford scientific 'clubb'; included here are Boyle, Petty and Hooke. It was at Oxford that the proto-Royal Society was formed; it was here also that Ward, Wilkins and Dalgarno began work on a philosophical language.

The shift from London to Oxford was not merely geographical; it was ideological as well. As the revolution progressed the reform motive of the vulgar Baconians grew more zealous and more radical. Their 'science' tended increasingly towards natural magic. With the regicide of Charles I, however, the counter-reaction set in; Hartlib and his circle were viewed with suspicion. Reform had opened a Pandora's box; science under the guise of natural magic bordered on the apocalyptic if not the heretical.[19] Serious science led by Wilkins retired to the ivory tower and the rationality of experimentation. The Oxford scientific group of which Wilkins was the prime mover became the locus of the scientific elite and of moderate social values. If the formation of the Royal Society marks the institutionalization and legitimization of the new theoretical science and the triumph over natural magic, it is not insignificant that Wilkins' *Essay*, which marks the culmination of the universal language movement was published under its auspices.

ii. Writing systems and real characters

After Bacon, the very earliest proposals for a real character came sporadically and mainly from France. Little is known about the circumstances which surround their composition, but their purpose is clearly utilitarian. The first was by Jean Douet in a tract entitled *Proposition présentée au Roy d'une escriture universelle admirable pour ses effets très-utile et necessaire à tous les hommes de la terre* (Paris, 1627). Douet's ideas for a universal character evolved from work he had begun in 1621 on codes and ciphers. This had led him to investigate Ciceronian and Tyrronian notes (shorthand), Egyptian hieroglyphics and Chinese writing, from which he conceived the idea of a universal character. What Douet envisioned there was 'a system of signs that could be read off by everyone in his own language, suggesting that the principle was that of assigning a character or characters to words in different languages that have the same meaning'.[20]

Sometime around 1629, another Frenchman, Des Vallées, devised a proposal which was brought to the attention of Descartes.[21] This occasioned Descartes' remarks on a universal language, about which we will have more to say below. Des Vallées had in mind a language which would consist of a dictionary of words considered to be primitive in, for example, French, as well as the synonyms for these words and their counterparts in other languages. Thus, the language might list *aimer* with its correlates, *amare*, *philein*, etc. There would be a single ideographic symbol created for each primitive word so that the same symbol would represent the French *aimer*, the Greek *philein*, and so on.[22]

Another French project often mentioned is the alphabet of Jean Le Maire, the idea for which originated in the 1620s. Little is known of Le Maire's specific proposals, but Descartes' colleague, Mersenne, reported to Hartlib's friend Theodore Haak that Le Maire promised a universal alphabet which makes it possible to communicate with any nation, including the Chinese, without having any advance knowledge of their language.[23]

In 1626 a Frenchman, D. P. Champagnolles, was corresponding with Philip Kinder in England about a universal character, the nature of which remains a mystery. Kinder himself devised a scheme he considered far simpler than the one of Champagnolles. In Kinder's scheme, the learner 'let him be but an ordinary adedarian may read and write within two hours space...the manner is easie, expedite for the succeeding numbers and row of Alphabet coupled hand in hand, is a true method for an artificial local memory, which is most capacious, most retentive'.[24] Kinder reveals to his correspondent, Beveridge, that his scheme is based on numbers: 'Thank Pythagoras that I have slept in's lap.'[25] In addition there are hieroglyphical marks 'borrowed from the Aegyptians', for signifying certain grammatical terminations.

In 1635 Champagnolles was still known to be in London perfecting his character.[26] In 1639 Hartlib's diaries mention that he had one book of the *Odyssey* printed in the character but 'does not know what is become of it'.[27] Champagnolles guarded his secret carefully and it was not until several years later, after his death, that the scheme was made known. In 1650 Ward and Petty obtained knowledge of it from Champagnolles' widow who was asking £100 for the scheme. Petty, with Ward's agreement, would 'not give six pence for it'.[28] Hartlib's journals for 1650 give a small glimpse of the scheme: 'But as for Champagnolles there are only some books of Homer done in the same character with a clavis upon it. The terminations or grammatical part are made by pricks in the following figure 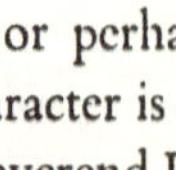 which is added to every word.'[29] The principle of Champagnolles' scheme (perhaps also of Kinder's) appears to be the abacus principle adopted by Gaspar Schott which we will treat later. The squares and the marks on the table signify numbers; the numbers probably each refer to a letter, or perhaps to a numbered list of words and/or terminations so that the character is an elaborate code.

Another projector of the 1630s was a Reverend Johnson of Ireland, whom we know through William Bedell, Bishop of Kilmore. Kilmore in turn was a friend of Hartlib, sharing with him irenic interests. Around 1633 Bedell persuaded Johnson to 'compose an universal character...the conveniency of it being so great, and the thing itself so feasible, seeing we have universal mathematical characters, as the arithmetical figures, the geometrical, the astronomical of the planets, and of the 12 signs of the zodiak, etc....He styled his book *Wit-Spell*. I have heard that some part of it was printed; but the rebellion of 1641 prevented the finishing of it.'[30]

Hartlib knew of Johnson's project as we see in his diaries of 1639–42 and Mersenne refers to Johnson's 'hieroglyphics'.[31] Later the fate of Johnson's project was described: 'the Fryars in Anthony have torne it into pieces for heresy; And you know that hee had allmost his copper plates cutt and graven, and wee heare, that they have suffered Tinkers to stop kettles with them'.[32]

The first published universal character appeared in Bishop Godwin's posthumous *Man in the Moone* (1638), a character based on musical notation.[33] As he wanders about the moon, Godwin's hero discovers a language 'not so much of words and letters, as of tunes and uncouth sounds, that no letters can expresse. For you have few wordes but they signifie divers and severall things, and they are distinguished only by their tunes that are as it were sung in the utterance of them, yea many wordes there are consisting of tunes only, so that if they list they will utter their minds by tunes without words...this was Gonsales:'[34]

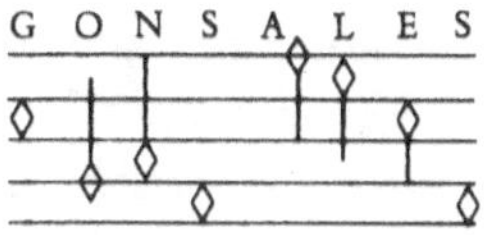

This lunar language put Godwin in mind of more earthly possibilities and in it he discerns 'means of framing a Language (and that easie soone to bee learned) as copious as any other in the world, consisting of tunes only...'[35]

This was not the first time that Godwin treated the subject of alternative kinds of communication. In *Nuncius inanimatus* (1629) he had discussed the possibility of communicating at a distance. While he keeps his invention secret, it appears that what he has in mind is a code rendered in light (somewhat like a semaphore) or in sound (like the musical language above). The lunar language of Domingo Gonsales would be simply a development of one of these alternatives.[36]

Godwin's whimsical language of tunes was quite possibly inspired by missionaries' reports that Chinese is a tonal language, and, insofar as it is simply a substitution of alphabetic letters by musical notes, it is nothing more than a cipher; indeed it may have been influenced by seventeenth-century works on cryptography.[37]

iii. Lexicons of primitive words: Wilkins, Comenius, Lodowyck, et al.

Godwin's works inspired Wilkins properly to invent the first real character in a treatise entitled *Mercury* (1641).[38] This is primarily an expansion of Godwin's *Nuncius inanimatus*.[39] After discussing a language of musical notation (borrowed clearly from the *Man in the Moone*) Wilkins describes other kinds of code signs, e.g. points and dots (like Morse code), geometrical figures, etc. This in turn suggested to him a universal and real character.

Like several of his predecessors, e.g. Bacon, Wilkins presupposes that what this real character represents is a lexicon of primitive words, words signifying the simple ideas out of which complex notions and terms are formed, 'To which purpose the Hebrew is the best pattern, because that language consists of fewest radicals.'[40] For each primitive word, there was to be an ideograph, to which must be added 'some particular marks to distinguish the cases, conjunctions, or other necessary variations of those derivatives that depend on it'.[41] Wilkins knew of the use of such a character in Chinese and Japanese (in which, he says, there are seven or eight thousand) but there were instances closer to home as well: many nations shared the same characters for numbers (Roman or Arabic) as well as signs for weights and measures ('philosophical number'), planets and constellations ('which characters were first invented by the ancient astrologers for the secrecy of them'), chemicals and minerals (e.g. △ for salt ammoniac), and musical notes.[42] These all worked by the same principle as the seventeenth century understood the Chinese ideograms to work: they were not alphabetical, and hence were not phonological; instead they referred directly to notions.

In the same year that Wilkins published *Mercury*, Comenius arrived in England to promote his educational schemes, and shortly thereafter wrote his first piece on universal language, ultimately published in *Via lucis*.[43] Whether this work of Comenius' is related to Wilkins' *Mercury* has been subject to debate. This began when De Mott argued that during his stay in England (1641–2) Comenius wrote his chapter on universal language which was then circulated in manuscript by Hartlib. De Mott argued that Wilkins was connected with Hartlib's circle at this time and that Comenius' manuscript is the source for Wilkins' scheme. Salmon has challenged this thesis on several grounds. First, Wilkins himself claimed that his source was Godwin's *Nuncius inanimatus*, about which there seems to be no doubt. Second, Salmon argues that there is no evidence that Wilkins was part of Hartlib's circle as early as 1641 (although as we have indicated through Theodore Haak this is not impossible). Third, she questions whether the chapter on the universal language was in fact written during Comenius' stay in London, whether it was written before Wilkins' book was published and whether it was left with Hartlib to be circulated. Salmon suggests that if there is any unique source, at least for Comenius and possibly for Wilkins, it is Marin Mersenne whose *Harmonie universelle* was published in 1636, a text we will discuss in the next chapter. As we have indicated, subsequent evidence has been adduced by Aarsleff to show that Wilkins was indebted to Mersenne, not only in *Mercury* but in all of his early writings.[44]

The path from Mersenne to Comenius and Wilkins is a tortuous, although not altogether impossible, one. On Mersenne's side, as we shall see, Mersenne and Descartes communicated about a universal language as early as 1629 and in 1634 Mersenne once again indicated his interest in the topic.[45]

We know also that during this period universal language was a subject discussed in Renaudaut's scientific conferences in Paris:

> the possibility of this Project appears, in that there is an order in nature, or, at least consequent to the very nature of things, according to which we may place, next after the Creator the created spiritual substances, then the corporeal (one after another, according to their dignity) particularly the corporeal according to their place, as the Heavens first, and in them the Stars, according to their dignity; the Earth and its Animals, the Sea and its Fishes, the Plants according to their Magnitudes, those which are equal therein according to their vertues, and other accidents: doing the same with Metals, Minerals... perfectly and imperfectly compounded by nature and by art; on with the Elements; everything in the world may be reduc'd and put into its right place. Whereby it is evident that all things have their order but also that he who learns them according to this order easily avoids Confusion, the mother of ignorance. It remains, now to find out an order of words too, which answers to that of things, the first to the first, and the second to the second; which order is so natural to them that Children make use of it to find out everything which they seek in Dictionaries and Alphabets according to the Order of the Alphabet.[46]

In 1636 Mersenne published *Harmonie universelle*, which set forth his scheme for a universal language. According to Mersenne, copies were sent to England to Cavendish and Digby, acquaintances of Hartlib.[47]

In England, we have Hartlib whose interest in a universal language dates back to Bishop Bedell's scheme in 1633. In 1635 he knew of the Frenchman Champagnolles' character. Up to the publication of Mersenne's *Harmonie* in 1636 there is no evidence of direct connection between Mersenne and the English. In 1638 however we hear that Johann Bisterfeld visited Mersenne in Paris after visiting Hartlib in London and possibly Haak in Leyden.[48] Bisterfeld was a partner of Alsted, colleague of Comenius, friend and correspondent of Hartlib and Haak. After his visit to Mersenne he wrote to Hartlib and

Haak mentioning a new and universal language.[49] By 1639 Haak was corresponding with Mersenne after Mersenne's opinion had been sought on the Pansophical schemes of Comenius. It is probable that either Hartlib or Sir William Boswell (a friend of Hartlib and correspondent of Mersenne) instigated the exchange.[50] The subject of the correspondence was primarily scientific information, but there are various connections between this correspondence and the universal language. Haak and Mersenne exchanged information about Le Maire, Johnson and Godwin's *Nuncius inanimatus*.[51] Mersenne wrote to Haak about the need for a new language, particularly one for botany, and he wrote to Comenius, via Haak, about his own language ideas in *Harmonie universelle*.[52] It is clear that Mersenne's ideas could have been the source for Comenius' reflections about the necessity of a universal language. And if Wilkins was associated with Haak at the time, or if Wilkins read *Harmonie universelle* independently, Mersenne's ideas could also have contributed to the development of *Mercury*. It is most likely that these early writings of Wilkins and Comenius were not directly related, but there is a good chance that, through Haak, they shared a common body of ideas deriving from Marin Mersenne.

Comenius' scheme is based on a combination of his Pansophical notions of a universal corpus of knowledge and his dissatisfaction with Latin and the other existing languages. What Comenius proposed as an alternative was 'establishing a language absolutely new, absolutely easy, absolutely rational, in brief a Pansophic language, the universal carrier of light'.[53] It was to be used in conjunction with his Pansophic ideas for universal schools and universal education.

However much a common language might ease the Confusion of Babel, its most important merit was that it could make a fresh start and, in so doing, it could set straight the relations between words and things. It could be:

> a universal antidote to confusion of thought. And it can only be that if its course is parallel with the course of things, that is, if it contains neither more nor fewer names than there are things; and joins words to words with the utmost precision as things are joined to each other, by constantly expressing the nature of the things with which it deals by the very sounds which it uses, and so presenting them to the mind...For the men of earlier ages, the authors of that language which has been handed down to us, were not so accurate in their investigation of things as to remark all their peculiar qualities and differentiae, and so to express them in exact and appropriate terms...[54]

This artificial language is to take the same shape as the basic Latin presented in Comenius' textbooks. It will have a lexicon, which is essentially an epitome of the notions basic to human knowledge. This is analogous to the early Latin vocabulary lists. There will be a morphology and syntax that is regular and consistent, free of redundancy and exception. It also requires regular derivation. (In the Latin grammars, it was the grammatical principles and paradigms of the regular cases that were to be presented first to the student.) The written form of this language will be rational in form as well as in matter, i.e. signs by which it is expressed will reflect not sounds, but the nature of things. In sum, 'every thing in [this] new language must be adapted to the exact and perfect representation of things...as things themselves present themselves in the same way to the senses of all men, they will make the language easy for men to understand and to copy'.[55]

The problem was, however, that the Pansophic taxonomy on which the language was to be based had not yet been created; that was the very point of scientific, Baconian investigation. Establishing the universal language must, therefore, wait 'till Pansophia has been made complete at least in the correct definition of the kinds, the ideas and the qualities of things. For it cannot be real and true unless it has been made apt to things evenly on all sides, nor can it be made apt in this sense until the foundations, the relative order, the mutual connections of things have been exactly revealed.'[56] Comenius himself would go no further; 'this duty we must leave to the College of Wise Men'.[57]

By the time that Comenius wrote this, the idea of a universal taxonomy of nature was widespread; so was an understanding of the problems involved in creating one. We see this from the remarks of Joachim Hubner found in Hartlib's journal. For all its claims, Hubner comments, Pansophia was 'nothing new': it 'seems to be nothing else but Rationarium Universalis [or] Logica Realis which does not consist so much in the combination of one term with another as in the combination and influence of one matter with another. So that one thing is applicable to divers in the same or some equal degrees of truth which he calls his Panharmonia.'[58] And again: Comenius 'scarce knows himself what he means by Pansophy. It may bee a Logicam Realem. His Metaphysica is the highest degree in it and should be the last of all set upon. For as yet it is impossible because of the want of particulars and indications that it can bee in any ways complete.'[59] 'Comenius is too much vaunting. It is no small matter to write Pansophia ad totum genus humanum. It is difficult enough if it be directed only for one country or Europe, how much more for the whole world. Comenius goes no further than Lesno and his own native and less polished countrys and Hungaria for whom it is good enough.'[60] Finally: 'his Orthographia requires mighty credulous men. For he supposes a world of things and takes them for granted. An accurate philosopher or sceptic would never admit of them. Such a book will be very acceptable to students which are insured to believe all things from their cradle.'[61]

Despite Hubner's objections, Comenius had indicated the direction that projects for a universal language were to take. Rather than being merely writing systems, i.e. real characters, they were to be nomenclatures based on a taxonomic arrangement of primitive words and notions, an arrangement that is all-encompassing, that identifies the most basic classes of things, that names and defines.

The projector who constructed the most detailed of the universal languages was Hartlib's protégé, Francis Lodowyck. Lodowyck can be said to be the most dedicated of the language projectors, if not the most successful. His language work began in the 1640s and when Hooke's diary ends in 1680 we still hear of him meeting to discuss the universal language. A merchant and a member of the Protestant refugee community in London, he was associated with Hartlib who published his major treatises, *A Common Writing* (1647) and *The Groundwork of a New and Perfect Language* (1652).[62] Later, in 1657, Lodowyck was a contributor to, as well as a critic of, Dalgarno's schemes and in the same year he wrote a manuscript entitled 'Of an universall reall caracter' which expanded the semantic ideas of his earlier works.[63] In his *Essay* (1668) Wilkins acknowledged Lodowyck's help in constructing the orthography of his philosophical language. In the 1670s he was a close associate of Hooke's circle, which carried on discussions of the universal language after Wilkins' death. In 1681 he became a member of the Royal Society probably through the efforts of Hooke and presented to them several papers on a universal phonetics and alphabet.[64]

As the title of his first work, *A Common Writing*, indicates, Lodowyck was originally interested in a universal or real character but the significance of his work lies in his proposals for a lexicon. Lodowyck begins his language by positing a set of radical words (to be compiled in a lexicon at some further date) to which a set of 'augmentations' or derivational affixes could be added. The radical words were English; ideas for the augmentations come mainly from Latin but there are also some from English, Dutch and German. In the *Common Writing* the radicals are divided into two groups: action and non-action words. Many of these can be 'contracted' or grouped together by virtue of four relations:

1. Analogicall, as *to see*, and *to know*.
2. Synonimicall, as *to lament, bewaile, bemoane*, &c.
3. Contradictionall, as *to curse, to bless*, &c.
4. In respect of the substance wherewith it is acted, as *to moisten, to wet, to wash, to dip, to besprinkle, to baptize;* the substance wherewith all these is acted, is *moisture*.[65]

In each of these four cases we see that there is one underlying notion, e.g. *moisten*, that is taken to be radical, or a simple term. There are then sets of related words, e.g. *moisten, wet, wash*, that are analyzed as complex terms deriving from the simple root. These complex terms are seen to be variations on a generic theme. Although this is not a taxonomy, it moves towards one: it analyzes and it distinguishes component parts, the two steps necessary before rearrangement in a taxonomic table.

In addition to these relations, Lodowyck also defines certain 'augmentations' by which derivatives may be made from the root word. In the *Common Writing* there are five that can be added to radical verbs:

1. Signifying a Custome of the action, for which the Latines have a particular denomination, as *Minitor.* i. *Frequenter minor*, which in English we would express, *to threaten often.*
2. A similation or imitation, as to *Grecize*, that is either in speech or otherwise, to imitate the Grecians: so to *Platonize*, that is, to imitate Plato either in his opinion, or style in writing, or otherwise.
3. An Inchoation, as in *senesco*, that is, I begin to be old.
4. A desire to the action, as likewise in the Latine Tongue is expressed, in *Esurio*, that is, I long or desire to eate.
5. A diminution, as in Latine, *Cantillo*, I sing low, or little.[66]

In his later *Groundwork*, Lodowyck added to this list. One might express a series of gradual differences as *to go*, *to trot*, *to run*, *to amble*. Negation can also be expressed.

The format is similar for the nouns. In the *Common Writing* there are six augmentations for radical nouns. These signify:

1. The actor	Drinker
2. That wherein, or wherewith is acted	Drink
3. The inclination	A drunkard
4. The abstract of the denominative adjective	Drunkenness
5. The act	The drinking
6. The place accustomary to the action	A drink-house[67]

In the *Groundwork*, Lodowyck divided the radical nouns into proper and common, the common being 'common derivatives from most verbs'.[68] These have nine augmentations which differ slightly from those in the *Common Writing:*

1. the action denominated	amor
2. that acteth	amator
3. on whom acted	amatus
4. inclination to action	amorosus
5. qualification to receive the action	amabilis
6. the excess of the action	amabundus
7. of or belonging to an action or thing	doctrinalis
8. the abstract of denomination	amorositas
9. the manner of being or acting	amorose[69]

The proper radical nouns are 'those names of substances, whether real or fictitious, whereby they are distinguished in their species or individuums'.[70] There are four augmentations for the 'specifical' proper nouns:

1. the instrument wherewith is chiefly acted as, from to See the Eye
2. the time of the action, the houre of Dining, Supping.
3. the place of action, as of drinking wine, a Tavern of beer or ale, an Alehouse
4. the substance of the action, Meat, Drink[71]

'Of an universall reall caracter' is a manuscript on the same subject dating around 1657 and is probably connected with Lodowyck's critique of Dalgarno's early work. More clearly than in his other works, Lodowyck is here concerned with the requisites of rational grammar or a philosophical language 'when we shall not be limited to this or that particular language as we find it, but shall be at liberty in our invention and choice, we shall hinder but what we may contrive such a grammar as respecting no particular language, but the genus or kind of language, may truly expresse things and their affections by certain short, constant rules, rectifying all those defects we complain of in ordinary language'.[72] What follows is an outline of possible derivatives and grammatical terminations many of which, like those in the *Common Writing* and *Groundwork* are taken from other languages, particularly, Latin, Greek and Dutch; but in addition to these there are many that could be considered 'semantic'.

In Lodowyck's universal language, the radical words are to be listed alphabetically in a lexicon and each is to be assigned a number in serial order as well as an ideographic character. The written character starts with a musical staff. The sign for the radical is written between the second and fourth lines; marks for inflections are written between the fourth and fifth lines and the marks for augmentations have set positions around the radix. In the *Common Writing* Lodowyck transcribed the first chapter of St John's Gospel into characters; 'and the light shineth' is transcribed as follows:[73]

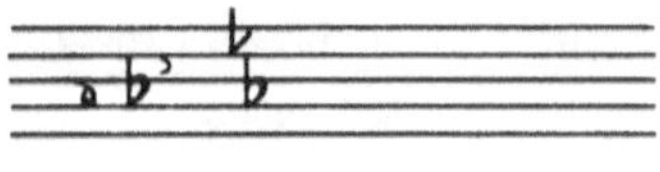

1. conjunction *and*
2. primitive verb *to light*
3. noun augmentation for 'that wherewith the thing is acted'
4. primitive verb *to light*
5. sign for 'relatival' augmentation, i.e. *to light* and *to shine* which are related insofar as *light* is the substance of both actions.

When this character is considered with the numbered words in the lexicon of radicals, we see that there are two alternative writing systems at work here: (1) ideograms for the words and (2) numbers for the *position* of the words in the lexicon. In the *Groundwork*, Lodowyck combined the two systems. He assigned ten ideographic characters to the numbers zero through nine. The characters for a word would be a combination of these signs which would express the numerical *place* of the word in the lexicon.

The simplification of the character in this manner permitted Lodowyck to create a phonological counterpart for his written system. Since he had only ten parts, he could assign to each a sound or a letter; the combination of those sounds or letters still expressed the numerical position of a concept or word in the lexicon. But, as Lodowyck explained, 'it will be prerequired to have ordained a certain number of single sounds...really such, and which may constantly keepe their stated sounds without alteration, out of which the foresaid Radixes should be worded. Thus might these Radixes be ordered and ranked, first, in their sounds in any known Language, thereto adjoyning the new invented words for the same.'[74]

In its simplicity, Lodowyck's language reveals not only the logic but the problems inherent in an artificially created language of this type, the most important of which was analyzing things so that they could be defined and appropriately named. Lodowyck himself did not press the point: 'the proper names of things to give them signification, is the work, we suppose, of a sound philosopher, who from the knowledge of things and their order in nature, should give them names accordingly, describing that in them by their name, by which in the naming they may be known'.[75]

Lodowyck's schemes were received with interest by members of the academic and scientific communities. The comments from Boyle are typical: 'If the design of the Real Character take effect, it will in good part make amends to mankind for what their pride lost them at the Tower of Babel. And truly, since our arithmetical characters are understood by all nations of Europe in the same way, though every several people express that comprehension of their own particular language, I can conceive no impossibility, that opposes doing that with words, that we see already done in numbers.'[76] But Boyle also had his reservations: 'only if the dictionary...do not overswell and disease it of a tympany'.[77]

The first scheme to incorporate a Comenian classification based on Latin root words appeared in Rome in 1653.[78] This was a broadsheet entitled *Arithmeticus nomenclator* by Pedro Bermudo, S.J. The scheme divides all things into forty-five general classes of Latin words, e.g. elements, heaven, the secular state, artists and artificers, affections, etc.[79] Under each class is a list of subclasses; for example, under Class XXX, Trees and Fruits, are listed: vitis, uva, oliva, malum, pyrum, persicum, pomum, prunum, etc.[80] The lexicon of words is to be arranged by number. Thus the first line of the Creed, 'Credo in Deum Patrem omnipotentem' is written in character as XXIX.4 XLII.8 III.1....XXXIII. 47 XL.23....(the dots following the Arabic numbers are signs for grammatical endings).[81]

In the early fifties two more projects are mentioned in Hartlib's diaries. In 1653 he refers to Potter, who is 'about a common language to be written in ordinary characters'.[82] In 1655 the diary reports that Benedict Skytte, a Swedish nobleman and Pansophical friend of Comenius, was 'in Sweden labouring mainly to make out the etymology or harmony of tongues and to make a universal language'.[83] In the meantime, Hartlib's friend the hermeticist Elias Ashmole extolled the virtues of the universal language: 'if this invention were but complete, Arts would arrive at a high perfection in a little space, and we might reckon on more time in the short account and measure of our days, to be employed in a substantial study of matter'.[84]

Although Lodowyck's language schemes show much greater depth and linguistic analysis, the first completed language scheme was Cave Beck's *Universal Character* (1657).[85] Beck was an Ipswich schoolmaster also known for his abilities in mathematics and cryptography. Perhaps because of his pedagogical experience, his concern in creating a universal language was not theoretical but practical, to provide a 'Mechanicall help for the unlearned or such as convers with unlearned strangers'.[86] He recommended it as a 'Pocket Mercury to Travaylors'.[87]

Beck's aim was to create a character which could be read off in any language. His scheme therefore provides about 4000 primitive English words and a grammatical apparatus based on Latin. Since the scheme is meant to be simple, there is little attempt to analyze words semantically. He introduces a few simple Latin-based affixes as derivational aids e.g. actor, causative, inceptive, etc.[88] In a separate section he lists 'the most unusual words which come into composition', for which he creates separate characters. These 'words' include for example *often, little, colour, disease, office, love, good, easy, wild.* Beck envisions a lexicon with entries reduced under philosophical heads, 'a small Treatise in the nature of Comenius' *Janua* set forth with the vulgar language on one side, and the Character on the other'.[89] In his present treatise, however, he contents himself with an alphabetical list numbered in serial order.

Beck sent his manuscript of the *Universal Character* to Wilkins who tried to dissuade him from publishing it. Among the projectors, it was Dalgarno who had the most appropriate comment for Beck's project: 'As for my own judgment if it can be taken for impartial, I find it nothing else, but an enigmatical way of writing the English language.'[90]

In the 1650s Wilkins and a group at Oxford were actively engaged in research on a language that would be not merely universal but philosophical, a subject we will treat in the next chapter. Among that group was George Dalgarno who was to write the first of the philosophical languages. This was not to come, however, before he first created a universal one.

Dalgarno joined the Oxford circle about 1655 or 1656. We know little of the man other than that he attended Marischal College of Aberdeen University and was a schoolmaster in Oxford for about thirty years.[91] In April 1657, Faustus Morstyn, a Pole residing in Oxford, wrote Hartlib that Dalgarno was helping him study shorthand. Morstyn also noted that Dalgarno was working on a scheme of his own consisting of a number of classes of things and symbols to represent them.[92] In the same month Hartlib recorded that 'Dr. Ward and Dr. Wallis are assisting a Scotchman to perfect his investigation of Real Characters.'[93] Wilkins and Ward, Dalgarno reports to Hartlib, had been appointed his 'examinators' and at this time Dalgarno presented his linguistic ideas to the Delegates of the University, of whom Wilkins was one.[94]

We know of these ideas through letters to Hartlib and Pell as well as through Dalgarno's publications. In the spring of 1657 he put out a brief advertisement, 'A New Discovery of the Universal Character'.[95] This was followed in November by publication of the full scheme, the 'Tables of the Universal Character'.[96]

In his preliminary broadsheet, 'A New Discovery', Dalgarno proclaimed the merits of his invention, the most important of which was that 'the exercise of it [shall] bring a great deal of clearness to the acts of the understanding, in grammatical, logical and metaphysical criticisms and shall ripe the judgements to discover the superfluity of some languages and the sterility of others'.[97]

To Hartlib, Dalgarno emphasized another merit of his scheme: besides being a common character applicable to all nations it is 'A compendious and contracted way of writing for any language, far exceeding all inventions of tachygraphy [sic; shorthand] in this nation for writing of the English language'.[98] When Hartlib received a specimen of the character transcribing St John's Gospel 16: 1–12 in July 1657, he found the character so 'compendious' he thought that the whole Bible could be printed in it.[99]

Structurally, there is not much in Dalgarno's 'Tables' that is new to universal language schemes. His language is to be based on a simplified vocabulary (a combination of Latin and English). These vocabulary words he considers to be the 'simple notions of language necessary for communication and common speech' minus superfluities of synonyms and ambiguities.[100] The primitive words are divided into radicals which express notions of things, and particles which connect and relate those notions.

The Table of Radicals is a compromise between the kind of alphabetical listing done by Lodowyck and Beck and the kind of categorizing done in the school grammars and logics. As a mnemonic aid, Dalgarno's tables are composed of thirteen stanzas of nonsense verse;

> Everyone of those periods are subdivided in seven lines, whereof every line contains six notions and three of those going always together by way of one sense. Again in every period the first line is a fictitious story; the second an illation by the particle (for). The third comes in... with an exception by the particle (but). The fourth brings in an inference from that by the particle (therefore) and the fifth renders a reason of that by the particle (because)... the two last lines (nevertheless and moreover) give a contradiction to what has been spoken before.[101]

When I *sit down* upon a *high place*, I'm *sick* with *light* and *heat*
For the *many thick moistures* do *open wide* my *empty* pores
But when I *sit* upon a *strong borrowed* horse, I *ride* and *run* most *swiftly*
Therefore if I can *purchase* this *courtesy* with *civility*, I *care* not the *hirer's barbarity*
Because I'm *persuaded* they are *wild villains*, *scornfully deceiving modest* men
Nevertheless, I *allow* their *frequent wrongs*, and will *encourage* them with *obliging exhortations*
Moreover, I'll *assist* them to *fight* against *robbers*, when I *have* my *long crooked* sword.[102]

Opposite each of these lines, the contraries of the primitive words are given; thus the contraries of line one are *rise up*, *low*, *displace*, *whole*, *dark*, *cold*.

While these tables are in no way philosophically methodical, Dalgarno reasoned that, in verse form, the stanzas could be easily memorized, 'memory being the chief thing here to be considered and which is the chief excellancy of this design above other languages'.[103]

If learning the root words in Dalgarno's language is a simple business, writing them out in character is a fairly complicated affair. Stanza number is indicated by an ideographic character. The number of the word within the line is indicated by the angle of a line placed around the axis of the character. The line number in which the word occurs is given by the position of the line on the axis. The smaller marks for particles (about two hundred) are placed in various positions around the radical. In all cases, what the sign refers to is the position of the word in question in the Tables.

As news of Dalgarno's scheme spread, Hartlib solicited comments and received opinions on it. Having discussed it with Dalgarno, Robert Wood was not wholly 'convinced' but regarded it 'as a much better way of short-writing than that currently in use'.[104] Sir Cheney Culpepper thought the scheme 'unsatisfactory because there was no analogy between symbol and concept' and because he disliked the written form of the particles.[105] Although these comprised nearly two-thirds of the language they were denoted simply by points and were liable to confusion. Later Culpepper reported that the scheme was received with 'scepticism in Oxford, where people think the plan too good to be credited'.[106] Boyle concluded that Dalgarno's 'Universal Character will certainly be far more stenographical than any short writing but whether it will bee or how far the Real Character to all nations that cannot bee yet determined'.[107]

Since the structure of Dalgarno's language was a good deal like that of Lodowyck's schemes, he received Lodowyck's approval although Lodowyck criticized the written form of the language. He argued, not unreasonably, that, if one wanted to indicate stanza, line and word number, one might as well use numbers, since these are ideographic anyhow.[108]

John Wallis, the noted grammarian and longtime friend of Wilkins, also criticized Dalgarno's work. He told Dalgarno, as well as Wilkins, that a universal character

> was certainly feasible in Nature...but that [he] did not think it likely to obtain in Practice. Because this Universal Character, must be in the nature of a New Language ...So that, For all Persons, to Learn his Character, and have all Books written in it; is the same thing as to Translate all Books into one Language, and to have this language learned by all. Which if it cannot be hoped, of any of the languages now in being... much less is it to be hoped for, in a New Language, now to be contrived...[109]

Development of a universal language was not limited to England. In the next decade various schemes were also developed on the Continent. Father Philippe Labbé's *Grammatica linguae universalis missionum et commerciorum* is of interest because it was known by Wilkins and by Leibniz.[110] First written about 1660, it is basically a simplification and regularization of Latin.

Johann Joachim Becher's work was a similar kind of simplified Latin dressed out in an ideographic character. Known throughout Europe primarily as a chemist, he published his *Character* (1661) during his period at the medical faculty at the University of Mainz.[111] Becher set up a lexicon of over 10,000 Latin words arranged in alphabetical order and numbered consecutively. These numbers were to be one form of the universal character. Accompanying the primary lexicon were decoding or vernacular lexicons which give vernacular words, their Latin equivalents and the numbers for the Latin words. Following the lexicon, Latin grammatical terminations were given in a list numbered from 1 to 74. In the character, numbers for grammatical parts were added to the numbers for the root words.

Becher worried that numbers limited the character to those familiar with the Arabic numbering system. To avoid this problem he invented a kind of line drawing which acted as a real character for the numbers:

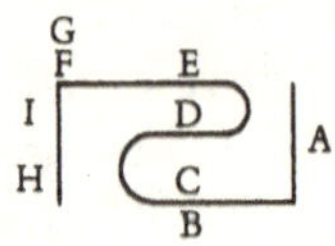

Dots and dashes placed in A, B, C and D represent units, tens, hundreds and thousands respectively, and these are given for the lexicon numbers of the root words. Marks in E, F, G, H and I represent numbers for inflections.[112]

Some years later, in reviewing Becher's character, Gaspar Schott constructed a simplified version.[113] Working on the same abacus principle as Becher, he uses a box for each word. The boxes are in fact duplicates of those found on seventeenth-century reckoning tables.[114] Dashes, which equal five, and dots, which equal one, go into the boxes to represent the number of the given word and its inflection in the lexicon:

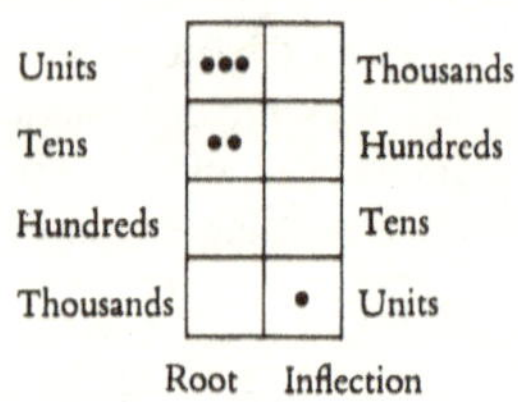

23.1 *equis* (nom. sg.)

Schott was a disciple of the scientist, Athanasius Kircher, who in 1663 also proposed a plan for a universal character.[115] Kircher's writings, as well as his interests, were numerous and varied. He dabbled in all the contemporary sciences, particularly chemistry, and in accord with the encyclopedic tendencies of the times wrote *Ars magna sciendi* based on the *ars combinatoria* of Lull.[116] His linguistic interests date back to the 1630s when he followed the French linguist Peiresc in studying hieroglyphics. He later wrote on Chinese. His works include also treatments of the Adamic language and the Curse of Babel.[117] *Polygraphia nova et universalis* (1663) was conceived as a secret cipher but it had other uses as well. Walter Pope writes Abraham Hill in November 1663 that Kircher 'has lately printed...a booke, which, if we may believe him is very curious; it is something like the universal character; it pretends to teach one of an indifferent capacity, in small time, to understand all languages, and to make himself understood to all. He intends to send it speedily to England to Sir Robert Moray...'[118]

Kircher's plan varied little from those that preceded it, particularly Becher's and Bermudo's, to which he may have had access. Like theirs, his was a universal character to accompany a simplified form of Latin. His basic Latin vocabulary of 1200 words is listed alphabetically and the equivalent words in Greek, Hebrew, Arabic, Italian, French, Spanish and German are given. The words are grouped into thirty-two sets of forty words each (the sets are simply the alphabetical listing broken up every forty words). At the end are special sets of words for countries, cities, periods of time, proper names, adverbs, prepositions and pronouns.

The character has three parts: Roman numerals to indicate the class of the word, Arabic numbers to indicate its position in the class (1 through 40) and characters for the grammatical endings, e.g. XXVII.36. NP (*Petrus*, nominative singular) XXIII.21 ∩E (*venit*, preterite singular).[119]

CHAPTER 6

Proposals for a philosophical language

i. Simple notions: Descartes, Mersenne, Kinner

The early universal language schemes provided neither an explicit analysis of language nor an explicit theory of nature to which language was related. To the extent that these schemes were anything more than writing systems they were, for the most part, simplified nomenclatures deriving from natural languages. Some were based on an assumption that nature (as represented in words) was constituted of certain basic sets of things (species), that these kinds of things had certain basic properties which defined or delimited them, and that they were arranged taxonomically. There were no explicit claims, however, that there was any relation between these nomenclatures and a model of nature.

A truly *philosophical* language must be based on an explicit theory and analysis of nature, and it must systematically reduplicate or reflect that analysis. It must represent, or be an image of, a (scientific) model of nature. It is this isomorphic relation between model and language that defines a philosophical language, as opposed to a real character or an artificial language which is merely universal. A real character or universal language is based on natural language. A philosophical language is based on nature.

The 'analysis' of nature which was ultimately adopted by the language projectors consisted of two parts, each predicated on a decontextualized view of nature. The first was the idea that all constituent things in nature can be analyzed in terms of a set of simple notions, natures or elements (instead of a set of primitive words); the second was that these constituent things are arranged in a taxonomic structure. As we have seen, Bacon's theory of nature included these two ideas: he considered bodies to be analyzable into a collection of simple natures and he assumed that these form things which fall into divisions or classes.[1] Bacon, however, did not specifically link this model of nature and the form of language, whether artificial or natural. He simply argued that the connection should be made.

Although it took the English projectors more than twenty years to reach this point, it was Descartes who first tied the analysis of things into simple elements to a proposal for a language, for an artificial, universal, philosophical language. These comments on a language scheme are found in a letter to Mersenne dated November 1629, replying to an advertisement of Des Vallées' project, which was, as we have seen, a character attached to a set of primitive words.[2]

Descartes argued that a universal language might be successfully developed if the whole idea of primitive words were abandoned. Rather than basing a universal language on radical words taken from existing languages, he would base it on the simple notions which constitute reality and on the laws of their combination or order. He argued that 'all the thoughts which can come into the human mind must be arranged in an order like the natural order of the numbers. In a single day one can learn to name every one of the infinite series of numbers, and thus to write infinitely many words in an unknown language. The same could be done for all other words necessary to express all the other things which fall within the purview of the human mind.'[3]

He noted, however, that the discovery of this philosophical language 'depends upon the true philosophy'. This philosophy provides a means by which men's thoughts can be separated out into 'clear and simple thoughts' which, in Descartes' opinion, is the 'great secret for acquiring true scientific knowledge'. If someone were to explain correctly the 'simple ideas of human imagination (out of which all human thoughts are compounded)' and if this explanation were generally received, only then would he 'dare hope for a universal language', a language which would 'assist men's judgement, representing matters so clearly that it would be almost impossible to go wrong'. Without this true philosophy, however, 'it is impossible to number and order all the thoughts of men'. Creating a universal language, in short, rests on the discovery of '*la vraye philosophie*'.[4]

Descartes' analysis of nature is not a taxonomic one but is rather a mathematical, atomistic one, related directly to issues he considered in the *Rules for the Direction of the Mind*, the earliest of his philosophical writings, believed to have been composed in 1628. Although it was never completed, the *Regulae* is a lengthy discussion of 'simple notions' as well as of their 'arrangement' in more complex forms.

In the *Regulae* Descartes elaborates a method for science, i.e. a method for the discovery of certain knowledge: 'Method consists entirely in the order and disposition of the objects towards which our mental vision must be directed if we would find out any truth. We shall comply with it exactly if we reduce involved and obscure propositions step by step to those that are simpler, and then starting with the intuitive apprehension of all those that are absolutely simple, attempt to ascend to the knowledge of all others by precisely similar steps.'[5]

This model is derived from arithmetic and geometry where, Descartes observed, the conclusions reached are absolutely self-evident and indubitable, and where complex problems can be resolved or solved by a long chain of very simple yet infallible steps. These mathematical sciences are based on simple, self-evident notions, and infallible logical sequences.[6] On analogy to arithmetic and geometry, Descartes hoped to create a universal mathematics, a method of similar simplicity and infallibility which would operate in all domains of science.

While it is clear what a 'simple notion' means in mathematics, it is not as immediately obvious what a 'simple notion' or a 'simple nature' is in the physical world. Descartes defines it as 'what is so clearly and distinctly conceived that we cannot divide it...it is not a further analysable object of knowledge and thought.'[7] These fall into three classes: purely intellectual notions (e.g. knowledge, ignorance, doubt, volition); purely corporeal notions (figure, extension, motion); and those which are common to both material and intellectual bodies (existence, duration, unity).[8]

What is understood by this is that corporeal things are differentiated quantitatively, i.e. by figure, extension and motion, all of which are measurable properties. This means that nature in its primary state is composed of particles of undifferentiated matter, i.e. of atoms. More to the point, it means that things cannot be explained qualitatively or as being essences or kinds divided into genera and species. It means that taxonomy is not a proper method of analyzing or representing the order of nature.

This distinction between the qualitative and the quantitative, between the atomistic and the generic, between the mathematical and the taxonomic was not properly understood by the linguistic successors of Descartes, viz. Ward, Wilkins and Dalgarno, just as it had not been properly understood by Bacon. Nor was Descartes' idea of 'order' properly understood. Descartes implies that in the process of analyzing the complex into successive stages of the simple, or in the reverse process of synthesizing the simple into units of greater complexity we follow the processes of subtraction and addition. A complex term is an implicatory series of simple terms, the simple term being contained in the more complex. Thus, as Joachim puts it, 'if X is a simpler term and Y more complex, Y will always contain X plus something else in addition. If X's character is precisely α, then the character of Y is $\alpha+\beta$; and if there is a third term Z, more relative and complex than Y, this can only mean, according to Descartes, that the character of Z is $(\alpha+\beta)+\gamma$.'[9] In the end, the more complex must be seen simply as an aggregate of simple natures, just as six is a fivefold addition of one and one.

This idea of order is clearly different from what is meant by order in taxonomy; there the relation of the complex (the genus) to the more simple (the species) is not an aggregate or a concatenation of elements. A species is a particular mode in which a genus exists; a man is not the combination of 'rational' and 'animal' as three is the combination of one and two but rather a man is a kind of animal, a kind of substance which is different from but in some ways similar to other kinds of substances. Genetic order cannot be reduced to mathematical order. The difference between these two is the difference between considering water, H_2O, as an aggregate of hydrogen and oxygen or considering it to have a nature which is different from its individual elements; or the difference between a sandwich and a piece of cheese put between two pieces of bread.[10] Descartes considered order to be one thing; the language projectors understood it to be another.

Once Descartes' linguistic proposals were made known to his colleague Marin Mersenne, Mersenne began working on his own version of a universal language. He was one of the early polymaths, almost singlehandedly responsible for establishing Paris as a major center of the new, scientific learning. Acting as the informal secretary of the French scientific and philosophical world, his correspondence reads like a *Who's Who* of the early seventeenth century, and it is not surprising to find that he was in contact with the English language projectors. Mersenne set forth his own proposals for a universal language in *Harmonie universelle*, begun in 1629 and published in 1636.[11] His point of departure was Descartes' contention that a new language (lexicon) would best be constructed from a small, finite number of elements. Mersenne then goes on to address himself to two points: one, that the vocabularies of existing languages are not sufficient to name all the (species of) things we need to name; and two, that it is mathematically possible to create the desired number of words from a set of simple notions.

Mersenne addressed the greater part of his discussion to this latter point in what forms a sort of lesson on combinatorial mathematics. He systematically explains and gives tables for the number of theoretically possible combinations one can derive, given one, two, or *x* number of elements. Applied to language, this demonstrates the number of 'words' one can derive from a number of consonants and vowels in various combinations. For example, if one were to compose all possible words from one vowel and two consonants (in three-letter words) out of a total of sixteen consonants and five vowels the total would be 3840; five vowels and two consonants in seven-letter words would make 16,800,000. If one were to use, as Mersenne proposes, the nineteen consonants and ten vowels of French, and if one varied the 'length' of a word, the number of possible combinations is astronomical.[12] Mersenne's point here is that it would be possible to construct enough names to number all the grains of sand and hence all the kinds of things. 'And if one says that God can always produce new individuals up to infinity, surpassing the number of words above', he adds, 'one can always add new words by increasing the number of their letters.'[13]

Mersenne's simple elements from which words are composed are qualitative properties of things, the only properties given us to know. As things are composed of properties, the names of things must include these properties and in his artificial words, each letter is to represent one of these simple notions. While he is not completely clear on this point, it appears that Mersenne assumes that the name expresses the species of the thing as well as the properties that differentiate one thing (species) from another.[14] To the extent that the language accomplishes this (by additional words, phrases, or whatever) his artificial words would constitute definitions of their referents:

If one takes the liberty to speculate about a universal language composed of all possible words whose root words are in sufficiently great number to provide different names for each property of all the species or all the individuals, and if one supposes that Adam...had perfect knowledge of all sciences and of the properties of each thing, one can imagine that he gave as many names to each species, e.g. to each animal as he recognized different properties...[15]

Mersenne's artificial words are composed of a number of phones, each of which represents a basic notion or a property. He argues that such a language might be built on a natural connection between sounds and things and concepts, 'in which the words or sounds had their signification naturally, so that men could understand the thoughts of others immediately as the words were pronounced, just as men understand that someone is happy when he laughs and is sad when he cries'.[16] Mersenne points out that there is some natural, onomatopoeic signification in natural languages (the examples he quotes are from classical poetry):

Many have noticed that there are proper letters for expressing softness or rudeness and other qualities of substance or actions and passions; the two vowels *a* and *o* are proper for signifying what is great and noble because an *a* is pronounced with a wide opening or beginning...*E* signifies delicate and subtle things, and is suitable for expressing mourning and sadness...*I* signifies thin and small things and is most proper for expressing humility. *O* expresses great passions and *u* signifies dark, hidden things.[17]

He runs through the consonants in a similar fashion but he is not overly optimistic about the feasibility of this principle for creating an artificial language.

Nevertheless, upon the completion of the *Harmonie universelle*, Mersenne wrote to the Orientalist Peiresc claiming that his language was such that a single reading could teach the philosophy or order upon which it was based. About this order, he reveals only that Peiresc's name is the 15,777,318,656th word in the language.[18]

It may well be that Haak received notice of this scheme from Mersenne and in turn reported it to Comenius and the Hartlib circle, as well as perhaps to Wilkins.[19] The person who saw the explicit application of these 'Cartesian' principles and who applied them, however, was Cyprian Kinner, who in turn transmitted them to the English projectors. He is the historical link between projects for a universal language and projects for a philosophical language. In his work the various influences of Bacon, Comenius, Descartes and Mersenne come together.

In his native Germany, Kinner was both a statesman and a pedagogue steeped in Baconian philosophy. He attended the gymnasium at Brieg with Hartlib and from 1634 to 1635 he collaborated with the encyclopedists, Alsted and Bisterfeld. In 1644 he formed a partnership with Comenius for producing school textbooks and, sometime between these two dates, he formulated plans for a botanical nomenclature of artificial words.

Like Comenius, Kinner was particularly concerned with systems of comprehensive knowledge, viz. with encyclopedic arrangements of things and notions, but Kinner and Comenius proved to be incompatible and their association ended in 1647. Kinner corresponded with Hartlib on various matters concerning educational reform, including an artificial language, and in 1648 Hartlib arranged for Petty to translate Kinner's *Continuation of Mr. John Amos Comenius School Endeavours*. Early in 1649 Kinner came to England in an attempt to find support for his projects; he died unexpectedly in May of that year.[20]

As was the case with Comenius, Kinner's pedagogical theories bear a direct relation to his linguistic proposals. He shared with Comenius a desire to base learning on a knowledge of things and their essences: 'That all Things Necessary to be known may be instilled into Learners...by the only meer help of sensible objects...For it propoundeth nothing, but what every man may with his own senses truly and properly conceive as an Ens or thing in being...'[21]

Similar to those of Comenius, and to a lesser extent to those of Bacon, Kinner's philosophical presuppositions are thoroughly Aristotelian, tinged now and then with a bit of Paracelsus. His methods of inquiry into the nature of things are essentially the classic (taxonomic) procedures of division and composition found in Aristotelian logic. His plan for the 'presentation' of sensible objects to students consists of a number of steps: the first inquires into the outward nature; the second analyzes things into their component parts; the third studies the ways in which the parts may be reassembled and/or combined. Once the natures, both inner and outer, are known, the foundation is laid for the arrangement of things in classes by genus and species.

The study of individual things begins with a description of (Paracelsian) 'Marks or Signatures, their Frame or Structure...the outward habit as well as the total particular of each...'[22] Thus, students learn to apprehend things 'distinctly, by some one or more differentiall characters'.[23] This is followed by rigorous comparison of things, that is to say, by comparison of the degree to which things agree or differ 'in their marks or signatures'. The frames or structures of things are treated 'analytically'; they are taken to pieces and divided 'into their Integral parts, as also things naturall into their parts, Essential and Accidental'.[24] After this, Kinner teaches 'the Virtues and operative Faculties...how by their signature and structure...to know their strengths and powers'.[25] The next step is to organize the likenesses and differences into classes of words, 'then hewing those classes till they may become fit to lye in a well ordered fabrick...'[26] Once the fabric or classification is established, Kinner indicates he will compose a dictionary of radical words and a directory that explains 'every tribe in the classes...and also the differences and other accidents of all words'.[27]

In his *Continuation*, Kinner did not extend his ideas on language learning to fit an artificial language, but his correspondence with Hartlib indicates that he understood the connections. He writes to Hartlib of a philosophical language written in both alphabetic letters and ideographs. Kinner's ideas date to around 1645 and he communicated them to Hartlib in June 1647.[28] He tells Hartlib that several years before, when he had been teaching botany, he had outlined a proposal for a botanical nomenclature.[29] In this plan, letters and syllables were to signify primary and secondary qualities of the plant named, and vowels were to signify the degree of these qualities. Additional syllables were to give further information: the second syllable, for example, was to denote the peculiar power or use of the plant, e.g. curative, preservative, nutritive, etc. The third syllable could indicate where the plant grows or how it is harvested.

Kinner had thought at first that this scheme was practical enough to extend beyond botanical things, for example, to elements, astronomical bodies, minerals, metals, animals and perhaps even to men and angels. He had outlined his ideas to Comenius who was 'interested' but nothing came of it. Kinner gave up the scheme thinking it 'would be too intricate and local to be satisfactory, since the characters of the Latin alphabet are more or less haphazard, can be distorted in many ways, and are not known to all peoples'.[30]

Having rejected alphabetic words, Kinner went on to devise a scheme based on ideographs. This character 'would be of a kind adapted to variations by additional marks on a scale from simple to complex; the simplest figure would be used for elements and the most complicated for animals and man himself. There would be as many separate figures as there were "primitive words"...a word is primitive only if it denotes a thing whose nature it is to be in some respect common to several things...'[31] That is to say, genera will be represented with primitive words.

What is interesting about Kinner's scheme is not only that he conceives of artificial words for things as being composed of signs of their simple elements, but that his model is perfectly explicit. As a colleague of Comenius he shared the master's notions of educating students about the nature of things. His program involved analysis of the outward nature of things, analysis of the structure and its component parts, assembling the parts, the things, and their kinds, into a system of order. This is the classic paradigm of taxonomy.

Wilkins' colleagues at Oxford were no doubt aware of Kinner's proposals and that Kinner was aware of their efforts at Oxford is known by entries in Hartlib's journal for 1650. There it is recorded that 'Mr. Ward is about another kind of more real and comprehensive characters. As Dr. Petty was trying to know the virtues besides the names of all herbes and plants, which Mr. Ward is studying to make universal to other things. Of this also see Dr. Kinn[er's] letters and Elucidar[ium].'[32] Petty had been interested in language and educational reform for quite a while. In 1648 he published *Advice to Mr. Hartlib about the Advancement of Learning* where he follows in the tradition of Bacon and Comenius recommending that education be reformed in the direction of the study of sensible things rather than of words.[33] He recommends that encyclopedias be compiled as textbooks which would eliminate the study of 'the nice, contentious and merely fantastical' and further, that students be taught to write 'swiftly and in real character'.[34] Its use, he observes, among other things would lessen the need for learning foreign languages. In another essay on studies, Petty recommends that between the ages of thirteen and fifteen, students be taught 'Universall grammar, first applied in the mother tongue, then the Latin tongue, and others in more use and service...'[35] Given his interests, it is not surprising that Petty translated

Kinner's *Continuation*. As far as his creating a botanical real character is concerned, it might be pointed out that although Petty spent time with Mersenne in Paris between 1643 and 1645 he did not take up his plans for a botanical nomenclature until 1650, the year after Kinner's visit.[36]

At this same time, Hartlib's friend Elias Ashmole was optimistic about this philosophical language: 'if some general Forms and Characters were invented (agreeing as neer to the natural quality, and conception of the Thing they are to signifie, as might be;) that (to men of all Languages) should universally express, whatsoever we are to deliver by writing; it would be a welcome benefit to Mankinde...Nor will this unity in Character seem impossible, if we consider there is in all men one first principle of Reason, one common interior Intelligence, and that originally there was but one Language.' What the project requires is the 'judicious direction of an able and general Linguist; and such a one that rightly understands the first and true impressions, which Nature hath stamped upon the things they would have signified by the Form'.[37]

There appeared at this time another, rather odd text that seems to be based on principles similar to Mersenne's and Kinner's. This was Sir Thomas Urquhart's *Ekskubalouron* (1652) which contained proposals for a universal language, and which was reprinted in *Logopandecteision* (1653).[38] Urquhart is best known as the English translator of Rabelais' *Gargantua and Pantagruel* (1653). His universal language scheme is rather Rabelaisian itself, the Epistle Dedicatorie addressed to 'No-Bodie'.

Two things are worth mentioning about Urquhart's scheme. First it is based on the principle that 'In the many thousands of words belonging to this language, there is not a letter which hath not a peculiar signification by itself.'[39] Presumably these 'letters' stand for simple elements.

Secondly, these primitive notions presumably form root words from which all other words are derived and it would appear that the root words are arranged taxonomically. Although Urquhart does not present any arrangement of these, he mentions, among others, faculties, science, art, trade, disciplines. The artificial words can express longitude, latitude, the position of the sun, moon, stars and planets, as well as the year, month, day, hour, fraction of the hour. For colors, the words will signify the 'proportion of light, shadow or darkness commixed in them'.[40] Finally, there are plants: 'words expressive of herbs represent unto us with what degree of cold, moisture, heat or driness they are qualified, together with some other property distinguishing them from other herbs'.[41]

Symbolic languages of this sort were not new to Urquhart. In 1645 he had published *Trissotetras*, a trigonometry text based on Napier's recent development of logarithms. Here he devised a notational system for formulas: E, for example stood for *side*, PRO for *proportion*, S for *sine* and O for *opposite angle; EPROSO* can be read as 'side proportional to the sine of the opposite angle'.[42]

Relatively little is known about Urquhart's intellectual life and hence about the source of his ideas for a universal language. He was, from his school days, interested in 'optical secrets, mysteries of natural philosophie...and ways to accomplish trigonomical calculations...'[43] A staunch Scottish royalist, he was knighted by Charles in London where he resided between 1639 and 1642. He spent many years abroad but the period between publication of *Trissotetras* and his universal language scheme, 1645 to 1650, appears to have been spent in Scotland. He was clearly knowledgable about developments in the universal language but from what source it is impossible to say.[44]

In the years immediately following Kinner's visit to England, there was a great deal of activity on the universal language: activity pursued both in London by Hartlib et al. and in Oxford by Wilkins and his associates. The Ward–Wilkins–Petty discussions at Oxford and their investigation of Champagnolles' and Johnson's characters took place in 1650, the same year as the attempt at a real character for botany. In 1652 Lodowyck published his *Groundwork*. In that year also, Urquhart's treatise was published, followed by its revision in the next year. In 1653 Hartlib's friends Potter and Skytte were working on a character. The same year saw the publication of Bermudo's tables and their possible dissemination in England. The most significant event however was the publication of Ward's ideas as a result of a debate with John Webster. While the debate laid down the principles upon which the philosophical languages were to be constructed, it also revealed the distance between the Hartlibian reformers and the university scientists.

ii. Oxford scientists: the Webster–Ward–Wilkins debate

John Webster is a good example of the reformative zeal of the more radical Puritan sectaries as well as a good example of the sectarian sympathies with the more mystical of the scientific movements and their affiliation with 'natural magic'.[45] He was a physician who had dabbled in alchemy with the Transylvanian chemist Hans Hunneades, and had studied with the alchemical mathematician, Oughtred (as had Ward).[46] During the wars, he was variously associated with the Levellers, Seekers, Familiasts, Grindletonians and the New Model Army. There is no evidence that he knew Hartlib although they had similar interests and Webster was clearly aware of the ideas coming from Hartlib's group, including those about a universal language.

Webster's tract, *Academiarum examen* (1654) was both a plea for the teaching of natural magic in the universities (on the grounds that this led to social and intellectual enlightenment) and a condemnation of existing curricula.[47] Having obvious familiarity with the works of Comenius, Webster included universal language among the desirable scientific innovations. He appears to have been unaware that, at the time of his writing, Wilkins' group was working on this project at Oxford.

In Webster's opinion, men have been misled by the tradition that takes the spoken word, and it only, as the theoretical foundation of human communication. Among the possible alternatives to the spoken word, Webster considers the felicity of ideographic characters, which are seen in 'Hieroglyphical, Emblematical, Symbolical and Cryptographical learning'.[48] He adduces the usual arguments for the practicality of such characters:

> who can be ignorant of the admirable, easie and compendius use of all sorts of Symbolisms that have but any insight into Algebraick Arithmetick, or have but slenderly consulted with the learned pieces of...Oughtrede, or the elaborate tracts of...Harrigon [sic]?
>
> ...the numerical notes, which we call figures and cyphers, the Planetary Characters, the marks for minerals, and many other things in Chymistry, though they be alwaies the same and vary not, yet are understood by all nations in Europe, and when they are read, every one pronounces them in their own Countrey's language and dialect.[49]

In this vein also Webster notices 'the wonderful and stupendous effects that Polygraphy, or Steganography produce'.[50] Coming closer to natural language, he finds an equivalent to these 'Symbolisms' in Chinese characters.

Webster felt that the real merit of a universal language was to be found in the relation it established between words and things. But, for Webster, these relations were those that were established in the original Adamic language of nature, relations in which words are the external or perceptible realizations of the internal essences they express.

Webster claimed that, in contrast to languages as we know them ('they are but dead characters'), this Adamic language consisted of 'signatures' stamped on things in the process of creation (an idea which derives from the German mystic Jacob Boehme and Paracelsus).[51] Nature, for Webster, is likened to a parchment, written over with living characters or signatures, expressing the mysteries and secrets of divine wisdom. The 'writing' of the signatures was both the creation and the naming of things. Here indeed was the very principle of logos for 'Adam, made in this image of God which is his external word, was made in the outspoken word, and so lived in, understood, and spoke the language of the father.'[52] Creation itself was the 'breaking forth this characteristical word'.[53] Created things are thus but symbols of divine notions. In the language of Eden, Adam's names were the verbal symbols of God's natural symbols:

> this Angelical and Paradisical language speaks and breathes forth those central mysteries that lay hid in heavenly magick, which was in that ineffable word that was with God, and lay wrapped up in the bosome of the eternal essence, wherein were hidden and involved in the way of a wonderful and inscrutable mystery, all the treasury of those ideal signatures, which were manifest and brought to light by the Peripherial expansion and evolution of the serviceable word, or outflowing fiat, and so became existent in the matrix or womb of that generative and faetiferous word, from whence sprung up the wonderful, numerous and various seminal natures, bearing forth the vive [sic] and true signatures of the divine and characteristical impressions...[54]

As Webster would have it, the universal language could recover and restore the Adamic language. All it need do was represent the signatures God had originally assigned to things.

It was Wilkins' colleague, the scientist and mathematician, Seth Ward, who took up the defense of academia and answered Webster. Ward had been brought to Oxford in 1649 by Wilkins (then Warden of Wadham College) as Savilian Professor of Astronomy and was associated with the Oxford virtuosi. He was not only familiar with the subject of universal language from his discussions with Wilkins but was at the time formulating concepts that were to provide the direction for future philosophical languages.

Ward's treatise, *Vindiciae academiarum*, the Preface of which was written by Wilkins, dismissed Webster out of hand: 'Webster doth give...the freest prospect of his depth and braine, in that canting Discourse about the language of nature, wherein he doth assent unto the highly illuminated fraternity of the Rosycrucians. In his large encomiums upon Jacob Behem...he may sufficiently convince that kind of credulous fanatic Reformer he is like to prove.'[55] Besides their general intolerance for Webster's mystical orientation, Ward and Wilkins were particularly annoyed by Webster's charge that the universities were doing nothing about the universal character.

As Ward indicates, he, along with several others, had long been intrigued by the prospects and promises of symbolization, particularly by universal characters. Not only was there the case of Chinese characters but mathematical symbols captured his attention as well. Prior to his tenure at Oxford, Ward had studied with the mathematician Oughtred. The beginnings of algebra date to this period and, in its early stages, one of the marvels of this new art was the substitution of letters for unknown quantities as well as the appearance of signs for the operations of addition, division, multiplication, etc. As he recounts the development of his ideas on the universal language, Ward indicates that he understood the possible symbolic applications suggested by algebra:

When I first fell from that verbose way of tradition of the Mathematicks, used by the Ancients...into the Symbolical way, invented by Vieta, advanced by Harriot, perfected by Mr. Oughtred, and Des Cartes: I was presently extreamly taken with it, finding by this meanes, that not only the substance of those vast Volumes might be brought into the compass of a sheet or two, but that the things thus reduced were more comprehensible and manageable; the labour of the braine much taken off, and a way layd open (by the various comparisons and applications of quantities) for invention and demonstration of infinite propositions with more ease than before...And I was put upon an earnest desire, that the same course might be taken into other things ...My first proposall was to find whether other things might not as well be designed by Symbols, and herein I was presently resolved that Symboles might be found for every thing and notion...So that an Universall Character might easily be made wherein all Nations might communicate together, just as they do in numbers and in species. And to effect this is indeed the designe of such as hitherto have done anything concerning an Universall Character.[56]

Ward went on to judge that such a design was indeed theoretically possible, but that it was hardly practical since 'the number of severall Characters will be almost infinite (at least as great as the number of primitive words in the most copious tongues....)'[57] In his *Essay*, Wilkins attests that it was Ward who showed him the mistake in attempts to frame 'such a Character, from a dictionary of Words, according to some particular Language, without reference to the Nature of things, and that common notion of them, wherein mankind does agree...'[58]

Carrying his research into mathematics a bit further, Ward found his way around this problem. Instead of treating words, and hence things, as single items or unified entities, he would break them into their less numerous simple elements and then use these simple elements as the base for a universal language. As he explains:

it did presently occurre to me, that by the helpe of Logick and Mathematicks [the universal character] might soon receive a mighty advantage, for all Discourses being resolved in sentences, those into words, words signifying either simple notions or being resolvable into simple notions, it is manifest, that if all the sorts of simple notions be found out, and have Symboles assigned to them, those will be extreamly few in respect of the other...the reason of their composition easily known, and the most compounded ones at once will be comprehended, and yet will represent to the very eye all the elements of their composition, and so deliver the natures of things: and exact discourses may be made demonstratively without any other paines then is used in the operations of specious Analytics.[59]

A 'word' or 'name' in Ward's language is therefore to consist of a simple notion (with its 'modal variations') or a combination of simple notions. Every simple notion is to be expressed by one syllable; a complex notion is expressed by as many syllables as the number of simple ideas it contains. While the determination of the 'simple notions' was left to the future, Ward hoped they might be within the range of the number of letters of the alphabet.

It is difficult to know how Ward meant his simple notions to be understood. His ideas are mathematical yet they were used equally well by Dalgarno and Wilkins in building taxonomic schemes. This confusion is similar to the one in physics between theories based on atoms and those based on the Aristotelian tradition of *minima naturalia*, the differences between which have been explained thus:

The *minima* of different substances are qualitatively different from one another, whereas atoms can differ only in size, shape and motion. In the theory of *minima*, the property of *shape* is of no importance, whereas in the seventeenth-century formulations of atomism, shape is often crucial to the production of physical qualities. Finally, in the atomistic doctrine all change in nature and all qualities derive from alterations in the motions and configurations of concretions of atoms, whereas in the *minima* theory, *minima* can act upon each other to produce internal qualitative alterations. In sum, the atomic theory admits of only matter in motion and mechanical interaction, but the *minima* doctrine permits interactions of a distinctly occult nature.[60]

In a similar fashion, it appears that at this point, what Ward meant by simple notions were principal genera; presumably the modal variations account for the differences by which genera are distinguished into species. Either atomistically or taxonomically, what Ward had in mind was a representation of an underlying order and the universal language is a representation of that. The idea was interesting enough for one contemporary to rank it with current advances in astronomy and 'optiques'. More to the point, it provided the necessary theory on which Dalgarno and Wilkins built their philosophical languages.

Once it is understood that an artificial language is to be a representation of things and is to be based on the observation of things, the nature of the things in question becomes critical. In the work of Descartes and Mersenne, followed by that of Kinner, Ward, Dalgarno and Wilkins, the results of objectification are carried one step further: things are conceived as abstract entities capable of being analyzed in terms of their component parts and in terms of the totality of their relations. Whatever terminology is used – simple notions, genus, species – the point is that what is under consideration is an abstract entity and its perceptible properties. This is roughly equivalent to the botanists' focus on structure, i.e. on visible, morphological criteria for the analysis and classification of the nature of things. At this point one enters the domain of specialized science and taxonomy.

CHAPTER 7

Schemes for a philosophical language

i. *The first: Dalgarno's* Ars signorum

The Restoration saw not only the creation of the Royal Society but the publication of the first full-length philosophical language, Dalgarno's *Ars signorum* (1661).[1] It took seven more years for the most extensive scheme, Wilkins' *Essay Towards a Real Character and a Philosophical Language*, to appear. There are many ways in which Dalgarno's and Wilkins' works resemble one another, the most important of which being the fact that both schemes rejected lexicons of root words from natural languages; both were ultimately based on some idea of common notions and simple elements. Nor should this be surprising; Dalgarno had been closely connected with Wilkins and Ward at Oxford, and it was differences of opinion concerning the disposition of the simple notions that caused each to go in his own direction.

Between the publication of the tables in 1657 and the publication of his full-length language scheme, Dalgarno's proposals underwent substantial modification. This had already begun when he wrote to Pell in 1657 that the printed Table of Radicals had been altered

> to make all notions whereof a certain number can be reduced to one head derivatives from the radical character of verbs and adverbs, making the ground of derivation some use or quality most certaine...We shall have a description of the thing, to wit, the genus of it in the title which represents what classes of things it belongs to and the property of the thing expressed in the radical from which it comes. This way I derive all the members of the body, diseases, all the species of nature, the young of living creatures, diminutives, arts, aggregata of every one of these. I have great companies brought together where ever the nature of the thing will bear to be reduced to one of those heads...[2]

At the same time Dalgarno expressed his reluctance to over-classify and over-conceptualize his scheme:

> many learned persons here [Oxford] advise me to order the simple notions in a philosophical method withal joining the memorative helps chiefflie. Dr. Wilkins

advises this and is taking pains himself to put them in predicamental order. This method I did choose at the beginning but thereafter deserted it for two reasons: first I found that all the memorative helps would not answer the philosophical method. Secondly, I feared if I did pretend a philosophical order, there would be quot capita tot sententiae among philosophers which might occasion alteration and so the chief intended end of the design to be lost.[3]

In the first half of 1658 Dalgarno advertised and solicited funds for his 'discovery of an Universal Character and a new Rational Language', a discovery which 'works a more real and lively apprehension of the thing treated of'.[4] A testimony signed in May 1658 by Lodowyck, Boyle and Wilkins, among others, claimed that two Cambridge students had mastered the language in a fortnight.[5] Writing to Hartlib in the same month, Dalgarno reports that Hartlib's friend Dr Worthington, Vice-Chancellor of Cambridge, had been very receptive.[6]

It appears that Dalgarno and Wilkins spent a good portion of 1658 and some of 1659 working in close contact to develop their language scheme. It is presumably during this time that Dalgarno was making the major alterations and revisions. But there is also indication that by this time Wilkins and Dalgarno were developing what were two different schemes. In 1657 Hartlib's colleague Oldenburg writes that 'If Mr. Dalgarno's papers speake only of his dessein for an universal character, without particularizing the manner thereof, I think it will be no news to me; and I have more confidence in Dr. Wilkinsius for the compassing it to usefulness, than in any other.'[7] In the fall of 1659, the two were separated by Wilkins' removal to Cambridge and by Dalgarno's preaching tour, but Dalgarno indicates that they may have been separated by more than just circumstance: writing to Hartlib in November 1659, Dalgarno complains 'I fear my too much freedome and openes while I conversed with men may be by some not made good use of...'[8] He refers specifically to Wilkins

who, if he publish anie thing he deales neither ingenuously nor justly with me and I have fear if he attempt anie thing on this subject he shal have smal credit of it for besides that all he wil doe wil be to discover another mans labors, this chiefly wil reflect that...he wil doe it with several material defects for albeit I did for almost a yeers tyme by dayly Convers with him open up the whole mysterie so farre as the veine of my invention had then gone yet since the tyme of my retirement I have discovered advantages which I perswade my[self] have not fallen under his consideration.[9]

Despite Dalgarno's fears, his work went sufficiently apace that in 1660 a petition was made to the King for support.[10] It is clear from this document that the language scheme he had in mind at this date is the one printed under the title *Ars signorum* in the following year.

The new directions taken by *Ars signorum* are best introduced by a contemporary who closely followed the evolution of Dalgarno's various language schemes. Dr Robert Plot writes that Dalgarno

> endeavouring to improve the Art of Shorthand beyond what others had done, by expressing the auxiliary Particles of the English Language, by distinct points and places about the radical or integral words, after this manner that 'tis done by prefixes and suffixes in the Hebrew; found at last that there was no way to distinguish the affixed points which he intended to be used really, from those used before in the common way of Short-hand (where not only the Capital Character, but points about it were Alphabetical) but by making the principal Character itself, to which they were to be accessory, not Alphabetical but real.
>
> Thus having formed Tables both of Integrals and Particles, to be expressed by single Characters, he perceived at length, that he was gone unawares further than ever he intended, having not only improved the Art of Short-hand, but also discovered a real Character equally applicable to all Languages: And after he had pursued the Design of a dumb Character a considerable time, at last he perceived that it would naturally resolve itself into a Language, having this Advantage over any Character, that we may use our known Alphabet to express it, whereas in a Character the figures must be new ...The last thing he attempted...was the reducing the Species of Natural Bodies to the rules of Art...from a few general words allowed to be radical, the names of the inferior species should be made off by composition, adding to the general and radical word, one, or sometimes more such Words taken from the Table of Accidents as might describe the intended Species, and difference it from all others...[11]

With the *Ars signorum*, Dalgarno had adopted Wilkins' idea of creating a language scheme on the basis of a classification of simple notions. To follow these principles meant to change significantly the direction taken by previous artificial language schemes. Dalgarno's new scheme does not begin with existing root words as did former ones: it begins with concepts and from these concepts artificial and philosophical words are created. These projects can be described as *a priori* ones: an inventory of non-linguistic items, viz. notions and things, will be set out in advance; the language itself will systematically reproduce this inventory in the linguistic form of the new words. A philosophical language, as Dalgarno understood it, presupposed an analysis of the order of nature. The outward and sensible signs of the language proper (words) are then but reflections or embodiments of this analysis. Thus Dalgarno argues: 'he who would impose names of things rationally, must first introduce into that chaos, the form, beauty and order of an ideal world existing in the mind, by a sort of logical creation'.[12] It follows that 'the grammarian ought to impose Names on things in accordance with the ideas and logical rules that are deduced from the external nature of the things themselves as they exist. This ordered series of things is usually called Praedicamentum.'[13]

In constructing his philosophical tables Dalgarno went to Aristotelian logic and taxonomy for his model. Things are categorized according to their genera (the most comprehensive genera being Aristotle's categories), species and differences, justifying this on the grounds that philosophers 'uncover the natures of things by careful scrutiny of similarities and differences, and thus, by collection and arrangement according to the method and order among them'.[14]

The purpose of the classification is twofold: not only does it introduce order into chaos, but at the same time a thing is classified, it is also defined. Thus, definition, one of the desiderata of a philosophical language, rests upon the pre-existing classification. 'All demonstration presupposes definition, definition presupposes genus and species; genus and species presuppose an ordered predicamental series... there is no definition or demonstration where there is no predicamental series.'[15]

Dalgarno then goes on to consider the way of constructing such a series:

> Either (1) one can investigate by means of logical analysis a certain number of primary and most simple Notions, out of which all other complex notions can be constructed. Indeed, this analytical approach leads to a deeper understanding of things. Men of utmost ingenuity believe that the number of simple Notions attained through strict logical analysis is less than the number of distinct articulable sounds attained through strict grammatical analysis. On the other hand (2) one can construct this series by reducing all notions of nature and art, complex as well as simple, to a straight predicamental line.[16]

For the construction of his tables Dalgarno chose the first alternative and, as we will see later, Wilkins chose the second. In the first method, the classification identifies primitive genera and 'species', i.e. primitive or simple notions (and the Latin words that represent them). These tables, for example, generate the notions or words *leguminous plant* and *smallest.* These simple notions or words can be combined to signify the complex notion *pea.* In the second alternative, the genera are divided and subdivided into smaller and smaller subgroups until further division is no longer possible. In this case, the category *leguminous plant* would be divided into its many and various kinds; *pea* would be a species of leguminous plant and a simple word would be formed for it.

Dalgarno chose the first alternative over the second for a number of reasons. First of all, it created a smaller number of primitive notions and hence of primitive words to be memorized in learning the language (about 1000). In a straight predicamental series there are too many infimae species – somewhere between 4000 and 10,000, although elsewhere he claims that the number might be infinite.[17]

In addition to practical reasons there were also linguistic ones: Dalgarno believed that the function of a universal language was to encode the common core of language and not its specialized varieties which include technical and scientific terms. As he noted in the earlier *Tables* 'This design being chiefly intended for common intercourse...it is not to be expected that it shall contain the numberless multitude of the several species of nature and words of art, & technick words of all trades, for this no language is able to do, nor no one man able to understand...'[18]

The philosophical reasons, however, became the most compelling. Dalgarno was concerned to identify the simple notions out of which (our ideas of) complex things are compounded. Each artificial word in the taxonomic tables is meant to represent such a simple notion.[19] Turned the other way around, this means that the tables do not include or classify complex beings or notions or their terms, as, e.g. they do not include peas but only the simple notion leguminous plants. As complex beings are 'made up' of a combination of simple elements, so their names are to be made up of a combination of simple words. The point is that Dalgarno believed his philosophical tables treat only simple elements; as he understood the idea behind Wilkins' tables, it violated philosophical principles by including complex things and notions.

Related to this is Dalgarno's understanding of the nature of (complex) things. As both Boyle and Locke were to argue, whatever 'essence' there might be, it is hidden, obscure and only imperfectly knowable, if at all. What we know of a thing is only a radical plurality or an aggregate of its characteristics not its form or essence.[20] Given this belief, logical classification is pointless since the method of logical division, i.e. division of a genus by the difference of a difference down to the indivisible species, assumes that such division will eventually reveal the total essence of the thing in question. If there is no form, or if it is impossible to know the form, there is no necessity for this kind of full classification. The difference between the method chosen by Dalgarno and that chosen by Wilkins is the difference between a conception of nature which is essential and genetic and one which is logical.

While Dalgarno disavowed the efficacy of classification for the understanding of complex things, this was still the means of 'logical analysis' he chose for identifying and defining simple notions in his philosophical tables. In his taxonomy, Dalgarno proposed seventeen genera, each of which is assigned a letter which will be used in constructing the artificial words. Each of these seventeen is broken down into intermediate genera and these in turn are divided into their 'species'.

The highest genus is Ens, Res. This is subdivided into three classes: Substances or concrete notions, Accidents, and Concrete beings combining substance and accidents.

The Substances or concrete notions are further subdivided into imperfect and perfect. The imperfect consist of corporeal substance (body) and spirit; the perfect consists of man, i.e. the composition of body and spirit. There are no further divisions possible in these categories.[21]

The Accidents are subdivided into seven subclasses and the Concrete beings combining substance and accident are subdivided into three classes. The taxonomic tables are concerned with the further classification of these. The outline of the major genera is given below:

A Ens
H   Substances
      Perfect
U   Composition of body and spirit
      Imperfect
O   Corporeal substance (body)
Y   Spirit
E   Accidents
B   Mathematical
D   General physical
G   Sensible
P   Sensitive
T   Rational
K   Political
S   Common
I   Concrete beings combining substance and accidents
M   Mathematical
N   Physical
F   Artificial[22]

Because Dalgarno's taxonomy is not a wholly consistent one, it is difficult to show its general principles. Nevertheless, we can look at plants in the category of Physical concrete beings and roughly outline his levels of taxa.[23] First of all, there is an Aristotelian distinction between substance and accidents although it is modified by the composite category of Concrete beings. Concrete beings are divided into mathematical concretes, physical concretes, and artificial concretes. In the subcategory of physical concretes, there is an

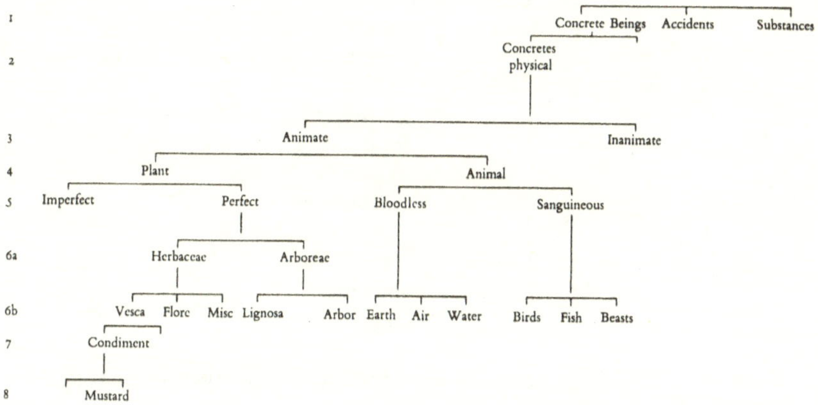

Fig. 5

implicit Aristotelian distinction between the animate and the inanimate; inanimate distinguishes the mineral world; animate beings comprise plants and animals, plants having vegetative power, animals having vegetative as well as sentient powers. These distinctions were a Renaissance–seventeenth-century commonplace.[24]

Plants are divided into the perfect and the imperfect as are animals (sanguineous/bloodless). This distinction derives from Aristotelian biology and again was commonplace. Perfect plants are subdivided into the large and small (arboreae/herbaceae).[25] Herbs are then classified as edible, flowering and miscellaneous. Each of these is in turn divided to produce the taxonomic tree shown in Fig. 5.

Although Dalgarno was not constructing a taxonomy for the same purposes as the botanists, what is interesting is that the degree of classification and the hierarchical structure of the classification is similar. Like the taxonomies created by the botanists, it reflects a combination of folk taxonomy with Aristotelian logical taxonomy. It is as though there were a taxonomic paradigm available to anyone who sought to analyze things. Specifically in this taxonomy the upper levels (1–3) are determined by logical distinctions. The categories of plant and animal (level 4) correspond to unique beginner in folk taxonomy; taken together levels 6a and 6b correspond to the life-form level.

This life-form level is divided into taxa we have called specialized. Herbs, for example, are subdivided into specialized taxa as follows (I omit the class of Miscellaneous):

6a	Life form	Herba
6b	Life form	Vesca
7	Specialized	bestiae vesca
		homini in folio
		in radice
		in fructu
		legumen
		frumentum
		condimentum
6b	Life form	Flore insignis
7	Specialized	bulbosa
		non bulbosa[26]

What we see in Dalgarno's addition of a level we call specialized is that his taxonomic structure reflects the contemporary state of the art – reflects the growing specialization of scientific knowledge.

The problem with Dalgarno's taxonomic structure is that it stops where it does – at the level of specialized taxa. These he felt elaborated the simple or primitive notions of things. The structure, however, does not include even the genera or infimas species of plants, and since it does not, it does not analyze, define, or name them.

Turning from the taxonomic structure to the nomenclature, what we see is that the language created by Dalgarno follows from the taxonomic tables. The same will hold true for Wilkins. Each in his own way reflects a desire to create a system of names that designate as well as define and also indicate the position of any given thing in the hierarchical system of nature. The artificial words and taxonomical tables create a double-layered nomenclature not unlike the one later developed by Linnaeus. The letters for the higher taxa name the genus. The 'speaker' of the language memorizes these. These letters/names function in the same way as the name *Arundo* functioned for Linnaeus in *Arundo calamogrostis* and *Arundo arenaria:* it signifies the genus. The letters/names for the specific epithets however do not name a class of objects directly. Rather like Linnaeus' trivial names (e.g. *calamogrostis*, *arenaria*) they indicate that each of the subclasses is different from the other and that each occupies a separate place on the taxonomic table. This is the same as Linnaeus' early practice of using numbers for specific epithets, e.g. Arundo 1, Arundo 2, etc. Dalgarno's specific epithets are codes which refer to places on the tables where the names of the objects are to be found. The names on the tables are the true, essential, diagnostic names.

These code letters/names are obviously not transparent or self-defining. Their function is to indicate that the groups of objects are dissimilar and occupy different, mutually exclusive places in the classification. In contrast, the diagnostic names found on the tables not only name the taxa but the name is at the same time a definition. They are exactly like the diagnostic names for plants found in pre-Linnaean botany.

In Dalgarno's language scheme, each of the seventeen main genera is signified by a letter. *N*, for example, designates Physical concrete beings. Each of the seventeen genera is divided into nine subclasses, each of which is assigned a letter. The letters in this case signify the serial order or the place of the subclass. The nine intermediate genera are signified by consonants:

1	2	3	4	5	6	7	8	9
m	n	f	b	d	g	p	t	k[27]

Like Linnaeus' binomials, these designatory names are linked to a parallel set of diagnostic names, i.e. the self-defining names listed on the tables. Turning to those, we find that

n – m signifies Concretum physicum; simplex
n – n signifies Concretum physicum; meteoron [sic]
n – f signifies Concretum physicum; minerale
n – b signifies Concretum physicum; vesca
n – d signifies Concretum physicum; flore insignis [sic]
n – g signifies Concretum physicum; lignosa
n – p signifies Concretum physicum; aerium volatile
n – t signifies Concretum physicum; aqueum
n – k signifies Concretum physicum; terrestre[28]

In each of these subclasses, the pattern is repeated: they are divided into seven 'species'. The letters indicate the number or serial order of each of these species:

1	2	3	4	5	6	7
a	h	e	i	o	y	u[29]

Turning to the tables, we find the diagnostic, self-defining names which state the essential properties of the species:

nab signifies Concretum physicum; vesca, bestiae vesca
nhb signifies Concretum physicum; vesca, homini in folio
neb signifies Concretum physicum; vesca, in radice
nib signifies Concretum physicum; vesca, in fructu
nob signifies Concretum physicum; vesca, legumen
nyb signifies Concretum physicum; vesca, frumentum
nub signifies Concretum physicum; vesca, condimentum[30]

An artificial name for a simple thing is composed of three to five letters/words similar to the way in which the English names *blueberry* and *blackberry* are composed of two words. Names for complex beings are to be composed

hay	concretum physicum	vesca	bestiae vesca									
nab	n	b	a									
mallow	concretum physicum	vesca	homini in folio	partes annuae	flos		accidens math.	affect. primae	longum			
nhblagban	n	b	h	a	g–l		b	a	n			
garlic	concretum physicum	vesca	in radice	qualitas sensibilis	sabor	pingue	partes annuae	folium		accidens math.	affect. primae	longum
nebghnagbana	n	b	e	g	h	n	a	g		b	a	n
onion	concretum physicum	vesca	in radice	qualitas sensibilis	sabor	pingue	partes annuae	folium		concretum math.	figura	cilandis
nebghnagmuba	n	b	e	g	h	n	a	g		m	b	u
cucumber	concretum physicum	vesca	in fructo	concretum math.	figura	cilandis						
nibmuba	n	b	i	m	b	u						
gourd	concretum physicum	vesca	in fructo	concretum math.	figura	conus						
nibmoba	n	b	i	m	b	o						
melon	concretum physicum	vesca	in fructo	accidens commune	gradus comp.	valde						
nibsuma	n	b	i	s	u	m						

beans	concretum physicum	vesca	legumen	accidens commune	gradus comp.	maxime	accidens math.	affect. primae.	rectum
nobsufbab	n	b	o	s	u	f	b	a	b
zizania	concretum physicum	vesca	legumen	accidens commune	gradus comp.	minime			
nobshuf	n	b	o	s	u	f–h			
peas	concretum physicum	vesca	legumen	accidens commune	gradus comp.	medium			
nobsluf	n	b	o	s	u	f–l			
wheat	concretum physicum	vesca	frumentum	accidens commune	perfectio	bonum			
nybsim	n	b	y	s	i	m			
barley	concretum physicum	vesca	frumentum	concretum artefact.	vibus	potus	victus		
nybflhm	n	b	y	f	l	h	m		
mustard	concretum physicum	vesca	condiment.	qualitas sensibilis	sabor	amarum			
nubgrhm	n	b	u	g	h	r–m			

Fig. 6

by periphrasis as they often are in natural language as in *Japanese maple tree, white water lily, baby lima bean*. One takes the simple root from the tables, e.g., Concretum physicum, vesca, in fructu (*nib*), and combines it with other words from the various tables of accidents, e.g. *cilandis, conus*. In effect, the periphrasis amounts to the predication of accidents to a concrete substance where the accidents act as the fundamental difference which defines the species. The resulting name is a definition.

Since Dalgarno's taxonomic tables do not provide the definitions for the genera and species of complex beings, Dalgarno must furnish a separate dictionary where these are defined. These lexicon definitions relate these items to their superordinate classes on the tables. In his lexicon, for example, Dalgarno gives the definitions for thirteen common genera of edible plants and he gives the artificial name for each. The name/definition for *mustard*, e.g. shows that it falls into the specialized class condimentum.[31] In this scheme, therefore, the functions of classifying and of defining are partially split apart. The tables only classify and name up to a point; the lexicon provides the rest of the definitions but not the classification. The diagnostic names for the thirteen genera of edible plants which Dalgarno lists in his lexicon are shown in Fig. 6.

In this scheme the functions of classifying and of defining are not fully brought together. Since the tables do not provide a full classification (which includes complex beings, and the genus and species levels of taxa) they do not provide a full set of definitions. These definitions for the genera and species were to be provided in the lexicon. Dalgarno realized that his very short lexicon was inadequate and acknowledged, as many had before him, that a full set of definitions would eventually have to be drawn up. What he did not perhaps realize was the inherent contradiction in his scheme. He dismissed classification as a means of analyzing complex beings; yet to make definitions of them presupposed a classification. The problem was that unless one taxonomized one could not define.

As Dalgarno's language scheme stands, definitions of complex beings are left up to the individual 'speaker' of the language. If he wanted to communicate about an avocado, for example, he would determine the proper genus and the appropriate characteristics (appropriate to him). He would then join the words for each of these. His definition is simultaneously his name for the avocado. The only problem is that it may not be anyone else's name. The name is self-defining relative to the speaker's individual taxonomy, but the taxonomizing and the defining are left to the individual. Variation between individuals can of course be expected. Dalgarno is faced here with the same problem which confronted the botanists: definitions and names vary from text to text and author to author. There was no common set of names fixed by common definitions and a common taxonomy. Except for the few names given in Dalgarno's lexicon, there is no common language.

Dalgarno's taxonomy is complete and detailed only through the level of specialized taxa. He does not connect the superordinate categories with the kinds recognized in nature. If we look at the plant names, we see that he classifies 32 kinds of plants, 44 including trees; the herbalist Bauhin had classified 6000. Because of this failure to connect, Dalgarno is unable to provide a complete analysis of (plant) things and their kinds. What lies in the gap between the higher and lower taxa is, of course, the whole field of specialized knowledge or science.

ii. Derivatives: Newton, Comenius

In between publication of Dalgarno's *Ars signorum* and publication of Wilkins' *Essay* there were two other language schemes based on a taxonomy and analysis of things into simple elements. Neither of them treated the taxonomy proper; rather they assumed its existence. They focused instead on simple notions which combined with the simple species to produce complex terms.

At the same time as publication of *Ars signorum* Isaac Newton was showing interest in the subject of a philosophical language. Newton's connection with the language projects and/or projectors in difficult to determine. He was a student at the King's School in Grantham in the years before 1661, the date of his admission to Trinity College. An early notebook of Newton's dated 1659–61 includes excerpts from Wilkins' *Mathematicall Magick*, passages in shorthand (Shelton's system, the same one used by Pepys) as well as a lengthy discussion of phonetics. The phonetic signs are quite similar to those used by Dalgarno in the 'Tables' of 1657 and by Lodowyck. The notebook also contains a list of some 2400 words arranged taxonomically under general heads beginning with Artes, Trades and Sciences and descending through Birds, Beasts, Instruments, down to Minerals.[32] Newton's universal language scheme occurs in a manuscript, the date of which is assumed to be 'just before or just after Newton's entry into Trinity College in 1661'.[33] This, it is believed, is connected with the phonetic material in the notebook.

According to the notes, Newton's scheme is for the creation of a 'general language' deduced not from existing languages but rather 'from the natures of things themselves'.[34] Since 'the frame of beings is made up of several substances (bodies and spirits) each of which are in this or that state or condition

...the use of language is that one man may signify to another in what state any substance is, hath been, shall be, may be, should be, is wished to be, is commended to be, etc'.[35]

For the 'names of things' Newton directs that 'In each Language gather an Alphabetical table of all substances (as Angel, house, I, thou, my thing, thy thing, this thing, that thing, good thing, great thing, etc.).' For each of these there is to be a name constructed in the universal language. In the process of this naming, a classification is made: 'It may be convenient to let the same sorte of things begin with the same letter, As of Instruments with s, beasts with t, the Souls passions with b, etc.'[36]

While assuming the taxonomic base, Newton then proposes several conjugations or derivational paradigms for the formation of complex terms. These include: '1. the degree to which an attribute may be ascribed to the given "thing"...2. The number of "things" involved. 3. The case or state in which the "thing" is. 4. The time when the given "thing" is or was in the specified state. 5. The quantity and number of times, i.e., how long and how often the given "thing" is or was in its case.'[37] These conjugations, as we see below, are strikingly reminiscent of Dalgarno's category of Common accidents which includes subdivisions covering causa, modi existendi, modi agendi, gradus comparationis, relatio numeri.

Newton's first set of modifications indicates the degree to which a substance 'is capable of being indued with some forme', e.g. indefinitely, exceeding much, etc. Also root words are capable of being quantified as well as qualified: another modification includes numbers like infinite, very many, many, pretty many, few, very few, one or more, one, all, any, some, any one, some one, none, but one, etc. A third conjugation gives the time when a thing is in its case and includes about three dozen possibilities for duration and the number of times.[38] The most interesting conjugation indicates the 'case or state' in which a thing is. The example Newton uses is the root word 'painter'. Terminations are added to this to indicate that any thing may be:

v	1. the remote or efficient causes or agents	the hire or persuasion moving him to paint
f	2. the mediating cause or instrument	the cloth, pencils, coloured liquors and other painting instruments
r	3. an action; or other praedicate of the subject	his painting action
d	4. the matter, patient, or correlate, or object	the thing whose picture he paints
rd	5. the forme of that matter, patient, or correlate	the form or abstract nature of the thing deciphered

k	6.	the terminus a quo and ex quo	a blanke, as unpainted cloth or paper
th	7.	the medium per quod	the unperfect picture
g	8.	the terminus ad quem and in quem and versus quem	the picture perfected
rt	9.	the forme produced, or form of the effect	the forms of the picture, as colour thus and thus disposed
t	10.	the effect	the picture perfected
b	11.	the end designed	the ends as to please the fantasy
p	12.	the end obtained	as to perpetuate the memory or the person painted
rl	13.	equally partaking of the common praedicate with the subject	
rm	14.	partaking of the praedicate more than the subject	
rn	15.	partaking of it less than the subject	
rk	16.	their difference or excess in one above the other in that praedicate	
rq	17.	the sum of all their participations in the praedicate	
rp	18.	so much the more partaking of the praedicate by how much the less the subject doth. et e contra.	
rt	19.	so much more partaking of the praedicate by how much more the subject doth partake of it. et e contra.[39]	

Another figure who is known to have made a contribution to a philosophical language at this time is Comenius in *Panglottia* (c. 1665–6).[40] We know that Comenius was aware of the work being done on universal language. In 1657 and 1658 he corresponded with Dalgarno and Hartlib about Dalgarno's schemes (which he considered 'wonderful, if they last').[41] The *Panglottia* indicates that he was familiar with the *Ars signorum*, as well as with Kircher's *Polygraphia*. While he does not seem to have known Wilkins' *Essay* first hand, his correspondence shows that he followed the preparation of it with interest.[42] His scheme furthermore quite clearly adopts some of Mersenne's proposals as presented in the *Harmonie universelle*.

Comenius is concerned primarily with constructing a nomenclature in which there would be one name for each species of things. His ideas on the construction of these names follows Dalgarno's rather closely. He argued that each letter or sound was to have signification, and that the word was to be self-defining. He was also, however, attracted to Mersenne's idea that sounds have a natural, onomatopoeic signification (they signify simple notions) and he made some tentative proposals based on this. One might say, for example, that *A* indicates large things, *I* something small and delicate, that *O* expresses roundness, brightness and universality; *L* is expressive of soft things while *M* signifies haziness, and so on.[43] These can be combined so that if *K* could be taken to signify gross size and *U* is immobility, then *KU* might stand for earth, massive element without its own movement.[44] Comenius realizes, however, that this onomatopoeic principle can lead to confusion and conflict and is willing to settle in other instances for signification by convention.

In Comenius' proposals, as in Dalgarno's, things are to be classified into their major genera. The names for these, constructed from significant self-defining letters as seen above, number about two or three hundred, but there is no taxonomy presented. Comenius is concerned rather with the grammatical requirements of the language. The most interesting feature of this grammar is a set of derivational affixes (classified as prepositions) which can be added to root words: *A*, privation; *E*, minor defect; *I*, diminution; *U*, increase, exaggeration; *O*, something excessive or universal.

Radical					
lus	alus	elus	ilus	ulus	olus
light	dark	a little light	shadow	blinding light	splendor
mel	amel	emel	imel	umel	omel
speaker	dumb	stutterer	babbler	shouter	tattler
lal	alal	elal	ilal	ulal	ilal
to speak	to be quiet	tell a story	whisper	shout	talk a lot
bar	abar	ebar	ibar	obar	
where	nowhere	somewhere	hardly anywhere	everywhere[45]	

CHAPTER 8

The culmination, aftermath and end

i. The final stage of development: Wilkins' Essay

It is in Wilkins' *Essay Towards a Real Character and Philosophical Language* that the language schemes finally become fully taxonomic, and language, just as nature, is analyzed and organized as a taxonomic structure. It is with the *Essay* also that we see gathered together the various lines of taxonomic endeavor – the taxonomy of the folk and of their language, the taxonomy of the Great Chain, the taxonomy of Aristotelianism, the taxonomy of the logicians, the taxonomy of the naturalists and botanists and the taxonomy of the language projectors. The *Essay* stands as a monument to that period in the first half of the seventeenth century when western Europe entered a new period of development which we call specialized, scientific, technical or (early) modern – a period brought on by an information explosion and new ways of knowing.

The *Essay* is the most comprehensive attempt of the century to readjust the taxonomic system: to expand folk taxonomy (and language) so that it incorporates and reflects the new knowledge we call specialized. It is perfectly appropriate that the most important pre-Linnaean botanist, John Ray, contributed to this project. It is not insignificant that Wilkins sought contributions from others considered to be 'experts'. Nor is the interest of other, more mathematically and mechanically inclined experts inexplicable. In the chaos created by the new learning, all of these specialists were seeking to find or to establish the order of things.

Wilkins' *Essay* had been in the works since the mid-1650s 'In the time of that daily and intimate converse which I injoyed, with that excellent Person Dr. Seth Ward [who] had more deeply considered [the universal language] than any person that I knew.'[1] According to Wilkins, by the time he had written the *Vindiciae*, Ward had made 'some considerable preparations, towards the Design...which if his other necessary imployments would have permitted him to have prosecuted, would without doubt, long ere this, have been advanced to as great a Perfection as the first Essay in so difficult a matter could have attained'.[2]

Ward's contribution at this time had been to demonstrate to Wilkins the mistaken nature of previous attempts at a universal language, viz. 'they did propose to themselves the framing of such a Character, from a Dictionary of Words, according to some particular Language, without reference to the nature of things, and that common Notion of them, wherein Mankind does agree, which must chiefly be respected, before any attempt of this nature could signifie any thing, as to the main end of it'.[3]

In the second half of the 1650s, while he was assisting Dalgarno, Wilkins tried his hand at creating a character,

> from a desire I had, to give some assistance to another person, who was willing to ingage in this design of framing a real Character, from the natural notions of things; for the helping of whom in so worthy an undertaking, I did offer to draw up for him, the Tables of Substance, or the species of Natural Bodies, reduced under their several Heads; which I did accordingly perform, much after the same Method, as they are hereafter set down...But he for whom I had done this, not liking this method, as being of too great a Compass, conceiving that he could sufficiently provide for all the chief Radicals, in a much briefer and more easy way, did not think fit to make use of these Tables. Upon which, being my self convinced, that this which I had begun, was the only course for the effecting of such a work...I resolved (as my leasure would permit) to go on with the other Tables of Accidents.[4]

Many years later Wallis recorded that:

> This enterprise of Mr. Dalgarno, gave occasion to Dr. Wilkins...to pursue the same design...Not that he did expect, this Real Character of his, and his Philosophical Language, should universally obtain; and all Books be translated into it: But, to shew the thing to be fesible; and divers Advantages which might arise thence, if it could so obtain.[5]

Plot records in detail Wilkins' objections to Dalgarno's scheme and the subsequent alterations he made in the conception of his own design:

> Thinking perhaps that [Dalgarno's] way of Composition would produce too long Words, or that the various Conceptions of Men must needs cause different Compositions and Descriptions of the same things, and thence unavoidably bring Misunderstandings (which yet Mr. Dalgarno thought might be avoided by stating of Notions and a Collection of Formula's) [Wilkins] did not approve of this Institution: declaring his Judgment to be, that all the Species of natural Bodies...ought to be provided for as Radicals, by a regular Enumeration of them according to the Theories of the Schools.[6]

Wilkins and Dalgarno went their separate ways geographically as well as theoretically. In 1659 Wilkins was appointed Master of Trinity College,

Cambridge. Less than a year later, with the Restoration in May 1660, Wilkins was forced to resign. He moved to London where he remained until his appointment as Bishop of Chester eight years later.[7] The formation of the Royal Society belongs to these years, its origin dating to November 1660 (the official charter was granted during the summer of 1662). Wilkins was very active in the Society and was elected its first Secretary at the time of the incorporation.[8] During the next six years he continued to work on his language project. By January 1666 Pepys reports that Wilkins was 'now upon finishing his discourse of a universal character'.[9] At the same time, as curator of the Royal Society, Hooke was arranging its collections 'under its several heads, according to the exact Method of the Ranks of all the Species of Nature, which has been compos'd by Dr. Wilkins and will shortly be published in his Universall Language'.[10]

While the *Essay* was not directly sponsored by the Society, it was clearly in line with Society activities both in providing a repository for data collection and in providing an encyclopedic classification of that data. The point is made rather nicely in Hooke's 'General Scheme or Idea of the present state of Natural Philosophy' (1666) where Hooke developed a notion of a repository of (information about) things, somewhat similar both to the Royal Society collections and to Wilkins' universal language.[11] While Hooke does not deal with the latter directly, he does present the scientific presuppositions upon which Wilkins' language was based. What we see in Hooke's treatise is the link between Bacon and Wilkins.

Hooke's intent in the 'General Scheme' was to set out an 'algebra' or method of scientific inquiry which would 'raise new axiomes or theories'. One starts with empirically collected information about real things and phenomena. This information, or collections, or 'matter of philosophical history', must be stored in an information bank or repository, and it is from these data that the new axioms or theories or general (mechanical) principles are to be derived. It is the repository that connects with the taxonomic tables of the philosophical language.

What Hooke proposes is that for more efficacious storage and/or memory, information is to be distributed into 'heads of inquiry', which need not be 'very nice or curious, they being in them laid up only in Heaps as it were, as in a Granary or Store-House; from thence afterwards to be transcribed, fitted, rang'd, and Tabled...'[12]

Hooke then gives such a taxonomic table; it starts with the major division of artificial and natural things, goes on to divide the visible world into the celestial and terrestial, and moves into the usual subdivisions. These taxonomically determined 'heads of inquiry' provide the individual subjects about which histories are to be collected or written. There are fifty-six such histories or mini-taxonomies suggested in the 'General Scheme' among which, for example, are the History of Mushrooms, Mosses and Plants, Roots, etc.; the History of Shrubs and Trees; the History of Light & Darkness; the History of Artificial & Mechanical Operations e.g., conversant with Stones, viz. Masons, Stone-cutters, Statuaries, Sculptors, Architects, etc.; Crystal-cutters, Engravers in Stones, Jewellers, or Stone-setters.[13]

These histories are to include two kinds of information: 'either a description of the things Themselves, whereby Inquisitive Persons that are ignorant of them, may come to a more perfect Knowledge of them...Or such a Description of them as is only in order to the Use of Philosophical Inquiry, for the Invention of Causes, and for the finding out the ways and means Nature uses, and the Laws by which she is restrain'd in producing divers effects. And this last is that which is chiefly aim'd at in this Design...'[14] This last is also the format for a real definition.

Hooke links these Baconian tables or histories with a philosophical language by suggesting that they should be transcribed in symbols: 'for this...is of huge Use in the Prosecution of Ratiocination and Inquiry, and is a vast help to the Understanding and Memory, as in Geometrical Algebra, the expressing of many and very perplex Quantities by a few obvious and plain Symbols: And therefore 'twere to be wisht, that we could express the whole History in as few Letters or Characters as it has considerable Circumstances...'[15] Working out these tables and symbols was precisely what Wilkins was doing.

While Hooke was composing his 'General Scheme', Wilkins was completing his *Essay*. By September 1666 it had gone to press when the Great Fire 'destroyed all the impression that was wrought'.[16] In his original version for the enumeration of plants and animals Wilkins had intended to use Merrett's *Pinax rerum naturalium Britannicarum*. Taking this opportunity to begin again, Wilkins abandoned Merrett's work because it 'did not meet his expectations'.[17] (Ray criticized it as 'bungling'.) Wilkins then asked Francis Willoughby to undertake the project and to enlist Ray's help.[18] Ray spent the winter of 1666–7 'in giving what assistance [he] could to Dr. Wilkins in framing his tables of plants, quadrupeds, birds, fishes, etc. for the use of the universal character'.[19]

Among other contributors to the *Essay* was Pepys, then Secretary of the Navy, who lent 'some of my tables of naval matters, the names of rigging and the timbers about a ship...'[20] William Lloyd, who we are told, 'had studied the nature of language in generall and of ours in particular very exactly drew the Tables for Wilkins' Philosophical language and Reall Character' and composed the Alphabetical Dictionary at the end of the *Essay*.[21] In this he was assisted by Wilkins' patron, the Duke of Buckingham. Lodowyck helped with the Orthography.[22]

In November 1667 Pepys records that Wilkins was 'talking of the universal speech, of which he hath a book coming out...'[23] Birch records that on 26 March 1668 the Council licensed Wilkins' book.[24] On 8 May he presented it to the Royal Society.[25] Pepys bought a copy from his bookseller on 15 May.[26] On 28 May, John Collins reports that the universal character 'so well pleased the King that his Majestie is resolved to make him a Bishop upon the next opportunity...'[27] He did just that in the fall.

One of the most striking features of Wilkins' book is that it is very large – 454 folio pages in the original edition – its size being proportionate to the encyclopedic intentions of the author. One benefit of its abundance is that in the book Wilkins attempts to say everything there is to be said about an artificial language and in so doing spells out the philosophical principles upon which such a project is founded, principles that were more often assumed than explained. The very first of these (deriving as we have seen from Aristotle, Scholastic grammarians and contemporary universal grammarians) is that out mental concepts are a faithful reflection of the phenomena of nature and since words stand for concepts, language is (or can be) a similar reflection:

> As men do generally agree in the same Principle of Reason, so do they likewise agree in the same Internal Notion or Apprehension of things...That conceit which men have in their minds concerning a Horse or Tree, is the Notion or mental Image of that Beast, or natural thing, of such a nature, shape and use. The Names given to these in several Languages, are such arbitrary sounds or words, as Nations of men have agreed upon, either casually or designedly, to express their Mental notions of them. The Written word is the figure or picture of that Sound. So that if men should generally consent upon the same way or manner of Expression, as they do agree in the same Notion, we should then be freed from that Curse in the Confusion of Tongues, with all the unhappy consequences of it...[28]

If there exists one real world for all, it is knowable by all; all that is required is a thorough cataloging of the 'things' of that world:

> The first thing to be considered and enquired into is, Concerning a just Enumeration and description of such things or notions as are to have Marks or Names assigned to them. The chief Difficulty and Labour will be to contrive the Enumeration of things and notions, as that they may be full and adequate, without any Redundancy or Deficiency as to the Number of them, and regular as to their Place and Order.[29]

Since our mental notions are signs for things, and words are signs for those notions, once things and notions are identified, a name or word-sign can be assigned to each of them. As in the case of botanical names, however, it would be best if those names were at the same time definitions:

> But now if these Marks or Notes could be so contrived, as to have such a dependence upon, and relation to, one another, as might be suitable to the nature of the things and notions which they represented; and so likewise, if the Names of things could be so ordered, as to contain such a kind of affinity or opposition in their letters and sounds, as might be some way answerable to the nature of things which they signified; This would yet be a farther advantage superadded: by which, besides the best way of helping the Memory by natural Method, the Understanding likewise would be highly improved; and we should, by learning the Character and the Names of things, be instructed likewise in their Natures, the knowledge of both which ought to be conjoyned.
>
> For the accurate effecting of this, it would be necessary, that the Theory itself, upon which such a design were to be founded, should be exactly suited to the nature of things.[30]

The way in which Wilkins intends to accomplish this is through a scheme based on tables of things and notions, that is, a classification by genus, species and difference. He indicates that he

> shall first lay down a Scheme or Analysis of all the Genus's or more common heads of things belonging to this design; And then shew how each of these may be subdivided by its peculiar Differences... After which I shall proceed to enumerate the several Species belonging to each of these Differences, according to such an order and dependence amongst them, as may contribute to the defining of them, and determining their primary significations.[31]

The reason why Wilkins constructed his tables in this way was that by so doing he was able (as Dalgarno was not) to provide not only names but also definitions for all the species of things:

> The principal design aimed at in these Tables, is to give a sufficient enumeration of all such things and notions, as are to have names assigned to them, and withall so to contrive these as to their order, that the place of everything may contribute to a description of the nature of it. Denoting both the General and the particular head under which it is placed; and the Common difference whereby it is distinguished from other things of the same kind...
>
> It were likewise desirable to a perfect definition of each species that the immediate form which gives the particular essence to every thing might be expressed...[32]

As a scientist however Wilkins was concerned to point out that determining or knowing the essence was an empirical rather than an *a priori* matter. Although he assumes that the nature of things can be known, in the present state of the art, the nature of things is what scientific investigation is all about.

Hence these first tables and definitions are necessarily tentative and are approximations: 'but this form being a thing which men do not know, it cannot be expected that it should be described. And therefore in the stead of it, there is reason why men should be content with such a description by properties and circumstances, as may be sufficient to determine the primary sense of the thing defined.'[33]

Wilkins hoped that his divisions would be natural, i.e. that they would be based on essential characteristics since only in this way would a taxonomy constitute a true representation of reality. He knew at the same time, however, that knowledge of those essential properties and essential natures had yet to be determined. That part of his classification was 'artificial' he admitted. His avowed intent in publishing the tables in their present state was to show that if not yet fully realized at the present time, such a true representation was theoretically possible.[34] As with Bacon, it is important to note that Wilkins does not deny the validity of classification as a method of defining essences. All he denies is that the proper classification has been made.

The specialized taxonomy found in Wilkins' *Essay* is the most fully developed one of his time. His taxonomical tables had to map out all taxonomical levels from the highest to the lowest. In contrast to Dalgarno, Wilkins understood that a taxonomy was legitimate only if it was complete and if it related, in a series of hierarchical links, higher taxa to the genera and to the individuals, just as the botanists understood that it was necessary to find the links between the species recognized in nature and the higher taxa dictated by philosophical tradition. Only in this way can things be defined, and defined 'really'.

For our purposes, the most interesting aspect of these tables is the collaboration between Wilkins and the naturalist John Ray who drew up the taxonomies of plants and animals. It reveals the coalescence of linguistic and scientific needs and theories. That Wilkins could engage Ray and that Ray could work for Wilkins shows the parallelism of their mutual endeavors. They were each confronted by a proliferation of things and information about things that required some kind of methodization and they believed Aristotelian taxonomy was the proper method. They needed and wanted to create self-defining names which would mirror the taxonomical outline and provide a groundwork for (scientific) communication.

The general outlines of the taxonomy were provided by Wilkins as were most of the tables. He based his taxonomy on the Aristotelian categories and the fundamental distinction between substance and accidents. He divides things into 40 superordinate genera as the table in Fig. 7 shows.[35]

The tables manifest the tendency which we saw in Dalgarno and other

- General: namely those Universal notions, whether belonging more properly to
  - Things; called TRANSCENDENTAL
    - GENERAL I
    - RELATION MIXED II
    - RELATION OF ACTION III
  - Words; DISCOURSE IV
- Special; denoting either
  - CREATOR V
  - Creature; namely such things as were either created or concreated by God, not excluding several of those notions, which are framed by the minds of men, considered either
    - Collectively; WORLD VI
    - Distributively; according to the several kinds of Beings, whether such as do belong to
      - Substance
        - Inanimate; ELEMENT VII
        - Animate; considered according to their several
          - Species; whether
            - Vegetative
              - Imperfect; as Minerals
                - STONE VIII
                - METAL IX
              - Perfect; as Plant
                - HERB considered according to
                  - LEAF X
                  - FLOWER XI
                  - SEED-VESSEL XII
                - SHRUB XIII
                - TREE XIV

- Sensitive
  - EXANGUIOUS XV
  - Sanguineous
    - BIRD XVI
    - FISH XVII
    - BEAST XVIII
- Parts;
  - PECULIAR XIX
  - GENERAL XX
- Accident;
  - Quantity;
    - MAGNITUDE XXI
    - SPACE XXII
    - MEASURE XXIII
  - Quality; whether
    - NATURAL POWER XXIV
    - HABIT XXV
    - MANNERS XXVI
    - SENSIBLE QUALITY XXVII
    - SICKNESS XXVIII
  - Action
    - SPIRITUAL XXIX
    - CORPOREAL XXX
    - MOTION XXXI
    - OPERATION XXXII
  - Relation; whether more
    - Private
      - OECONOMICAL XXXIII
      - POSSESSIONS XXXIV
      - PROVISIONS XXXV
    - Publick
      - CIVIL XXXVI
      - JUDICIAL XXXVII
      - MILITARY XXXVIII
      - NAVY XXXIX
      - ECCLESIASTICAL XL

Fig. 7

taxonomical theorists to reconcile Aristotelian theory with folk classification: first the Aristotelian division into substance and accident followed by the division of substance into animate and inanimate. Inanimate is composed of the elements; animate of the vegetative and sensitive. Wilkins has a slightly different version in classifying vegetative into imperfect, which in this case includes minerals and metals, and perfect, which includes plants. The sensitive, i.e. animal, is divided into bloodless (imperfect) and sanguineous (perfect).

The higher levels of taxa combine Aristotelian divisions with the folk level of unique beginner. The middle levels which encompass herbs (3 kinds), shrubs, trees, insects, birds, fish and beasts equate with the folk level of life form. It is at this point that the expert Ray stepped in and took over from Wilkins.

When Ray agreed to undertake the tables for the *Essay* he was entering the middle part of his botanical career. He had begun in the herbalist tradition, categorizing plants by a number of traditional variables (use, etc.). These, however, he knew produced artificial groupings: 'though these heads may seem more facile and vulgar; yet they are not truly philosophical, but depend too much on the opinions and customs of several times and countries'.[36] He had written the catalogue of Cambridge plants which was alphabetical, i.e. which did not attempt to systematize, and he had gained familiarity with Jung's work in morphology and the possibilities it offered for classification. His most significant advance came during his trip to the Continent between 1663 and 1666. It is at this time that he learned of Cesalpino's logical–philosophical method. Wilkins' tables provided Ray with the opportunity to apply those Aristotelian theories to create an Aristotelian taxonomy. It remains only to be said that by virtue of his efforts Ray started down a track which led him ultimately to reject the theory upon which the classification was founded – a rejection which occurred concomitantly with the demise of projects for a universal language. That part of the story remains for a later section of this work.

At this point, Ray was after precise definitions of major genera and their species which, as he understood, were predicated on 'a well-ordered arrangement into classes, primary and subordinate'.[37] The crucial element in that arrangement was determining the essential characters or differences upon which the divisions were to be made.

The 'truly philosophical' method which Ray found was based on Cesalpino's idea that classification needs to proceed by virtue of essential characters and these essential characters are morphological variables which visibly represent the structure, the structure being determined by the essential functions of the plant (nutrition and reproduction). This philosophical method is to proceed by logical division, where the classes are distinguished on the basis of a single character. Such a classification will result in a set of definitions – statements or descriptions of the essential natures (characteristics) of plants.

Ray is aware of the problems involved in adopting this logical method, especially the problem of finding the differences which are essential and common but which are not accidental.

I design these tables to take notice only of the *chief families* of Plants, to which the others are to be reduced...I take notice only of that *Communis ratio*, which belongs to all the subordinate varieties of them, unless it be when there is no such common agreement belonging to all of them: or where several things are reduced under the same head, some for agreement in one Accident, and others for their agreement in some other Accident...[38]

As for the infimae species (which these tables will not list) 'these need not be particularly provided for, because the just number of them is not yet stated, every year producing new ones: And because they may as well be expressed Periphrastically here as in all other languages...'[39]

The first division in Ray's system is between plants which are 'Minute and Tender' (herbs) and those which are 'Large and Hard, being wooddy'; the lesser of these are shrubs, the larger are trees. This is Cesalpino's division, although Ray admits that the distinction between trees and shrubs does not seem to have 'any such distinct limits in nature'.[40]

As with Aristotle and Cesalpino, herbs are either perfect 'having all the essential parts of a plant', or imperfect 'which do either want or seem to want some of the more essential parts of Plants, viz. either Root, Stalk or Seed', that is to say, they lack parts related to nutrition and/or reproduction.[41]

Since he was unable to find one variable according to which the subsequent divisions could be made, Ray proposed several. Thus the perfect plants are divided into three groups in relation to either their leaves, flowers or seed vessels. The characters of fructification (flowers and seed vessels) are borrowed from Cesalpino, while characterization by leaf had long been one of the more obvious devices of the herbalist groupings.

Once he establishes basic life-form categories, Ray begins the real work of classifying plants, i.e. of elucidating the specialized taxa which will be tied together with the genera. The requirements of Wilkins' taxonomy dictated that each life form could be broken into no more than nine specialized taxa, and the specialized taxa were broken into no more than fifteen genera. These fifteen genera are listed in pairs in order for the taxonomy and written character to accommodate numerous specimens.

The taxonomic tables are the primary language of things; what remains is to give the taxonomic structure linguistic signs so that they can be referred to and discussed. In the language scheme, therefore, letters are associated with each of the superordinate categories. *G* signifies 'plants'. There are five kinds of plants each of which is signified by a vowel:

gα plants classed by leaves
ga plants classed by flowers
ge plants classed by seed vessels
gi plants, shrubs
go plants, trees

These names (*ga*, *ge*, etc.) are not diagnostic names since they are not self-defining. The tables however provide the definitions:

gα minute and tender plant classed by leaf
ga minute and tender plant classed by flower
ge minute and tender plant classed by seed vessel
gi lesser hard woody plant with several stems
go larger hard woody plant with single stem.

In each of these genera there are nine subclasses, called 'differences'. As in the case of Dalgarno's language, there is a letter which signifies the number or serial order of the subclass:

1	2	3	4	5	6	7	8	9
b	d	g	p	t	k	z	s	n

Gab therefore signifies 'plant, herb/flower, 1st difference' and so on.

In the tables a parallel set of diagnostic names are listed:

gab plant, herb/flower, stamineous
gad plant, herb/flower, foliaceous, not pappous
gag plant, herb/flower, pappous
gap plant, herb/flower, umbelliferous, broader leaf
gat plant, herb/flower, umbelliferous, narrow leaf
gak plant, herb/flower, verticillate, fruticose
gaz plant, herb/flower, verticillate, non-fruticose
gas plant, herb/flower, spicate
gan plant, herb/flower, button-like head

Each of these differences is divided further still into fifteen 'species'. Again there is a letter signifying the serial order of the species, one through fifteen. *Gapα* signifies 'plants, herb/flower, 4th difference, 1st species'. The diagnostic name on the tables is 'plant, herb/flower, umbelliferous, of broader leaves, with a more pleasant odor, more properly belonging to this tribe, with different shaped leaves'.[42] Below is Ray's classification for the one specialized category 'umbelliferous herbs' and Wilkins' artificial names. What is to be noticed is that, used in conjunction with the tables, the artificial name is fully self-defining. It refers to a class of objects and defines them at the same time.

- Herbs according to their flowers — *ga*
  - Stamineous
    - Having no leaf (I) — *gab*
    - Having leaves, being either of
      - Compound flowers, whether
        - Not pappous (II) — *gad*
        - Pappous (III) — *gag*
      - Simple flower... distinguished by the manner of the growing of the
        - Flower, into
          - Umbelliferous... distinguished into
            - Broader leaves (IV) — *gap*
              - Odorate, whether
                - More pleasant, in such kinds of plants as are
                  - More properly belonging to this tribe distinguished on account of
                    - Leaves, as to their
                      - Shapes, whether,
                        - Different shapes in the same plant
                          - Annis–Coriander (1) — *gapα*
                        - Winged, resembling
                          - Parsnip
                            - Bastard stone Parsley (2) — *gapa*
                          - Fearn
                            - Sweet Cicely–Wild Cicely (3) — *gape*
                        - Not winged
                          - Alexanders–Lovage (4) — *gapi*
                      - Colour, whether
                        - Pale green
                          - Angelica–Masterwort (5) — *gapo*
                        - Dark green
                          - Laserwort–Herb Frankincense (6) — *gapγ*
                    - Hairy tuft
                      - Sermountain (7) — *gapy*
                  - Less properly belonging to this tribe
                      - Valerian–Meadow Sweet (8) — *gapyi*
                - Less pleasant,
                  - Lesser
                    - Parsley–Smallage (9) — *gapyγ*
                  - Larger

Hercules–All-Heal (10)	*glapα*
Not odorate, distinguished by	
Roots	
Parsnip-Skirret (11)	*glapa*
Leaves whether	
Whole	
Thorough Wax–Hares Ear (12)	*glape*
Winged	
Burnet Saxifrage–Umbelliferous Eringo (13)	*glapi*
Different shapes	
Candy Alexander (14)	*glapo*
Place of Growth	
Cow Parsnip–Water Parsnip (15)	*glapγ*
Finer leaves (V)	*gat*
Verticillate...their leaves growing by pairs, whether	
Fruticose (VI)	*gak*
Not fruticose (VII)	*gas*
Spicate (VIII)	*gaz*
Seed (IX)	*gan*

While the tables generate root words, more specific words are to be created by derivation or periphrasis, that is, by adding a related 'circumstance' to the root. This is accomplished by combining root words with a kind of particle called 'transcendental'. These 'do circumstantiate words in respect to some Metaphysical notion; either by enlarging the acception of them to some more general signification, then doth belong to the restrained sense of their places; denoting a relation to some other Predicament of Genus, under which they are not originally placed'.[43]

Wilkins recognizes affixation as the traditional way of 'varying the sense of words'.[44] Several of the transcendental notions are suggested directly by Latin suffixes. Latin *-tim*, for example, meaning 'to proceed one by one' as in *seriatim* sets up the category of 'segregate'. The combination for 'enclosure and service' comes by way of several Latin endings. The division 'sepiment', the enclosure of places, derives from the Latin ending *-ile*, 'animal pen', as in *ovile*, 'sheepfold', but the category is then extended to other things like *moat*, an enclosure for water. This in turn suggests 'enclosures for humans', 'armaments', which are fitted to their bodies, and does not come from Latin endings. But the enclosures of house and room are once again Latin based. House, according to Wilkins, is signified by Latin *-ile*, *-ale*, *-arium*, as in *caseale and aviarium*. 'Intruments and vessels' (for operations and/containment) come from the Latin *-trum* and *-bulum* respectively as in *haustrum* and *infundibulum*. But these simple figures suggest another division of more complicated figures – 'jugaments and machins' (long extended forms and more complicated forms involving motion) – which are invented by Wilkins.

The Latin ending *-arius* which signifies the general concept of a person belonging to an action or a thing, is broken down by Wilkins into several more specialized categories: officer/artist and mechanic/merchant. Wilkins admits there is no close affinity in the first pair. But in the second pair, he feels merchant properly forms its own category by virtue of the English termination *-monger*, from the Latin *mango* ('seller'), as in *fishmonger*, *whoremonger*. Inceptive and frequentive come directly from Latin verb endings, *-sco* and *-ito*. The six degrees of qualities or actions were probably born of the very general Latin endings for augmentative and diminutive: *-osus* as in *aquosus*, and *-lo*, *-lus*, *-lum*, *-aster* as in *cantillo*, *homunculus*, *corpusculum*, *grammaticaster*. Finding degrees of magnitude to be handy, Wilkins extends the notion 'excessive–defective' which seems to have no formal counterpart in language. English terminations also play their part in determining some of these categories. The obvious case is *like*. The ending *-age* as in *herbage* suggests the category 'kind'. English has a causative particle in the archaic form *be-*, *besot*. Wilkins sees also certain forms of *dis-*, *mis-*, and *un-* as 'corruptive' as, for example, *discolour*, *misuse* and *unlucky*. The idea of the 'perfective' is directly attributable to Greek *ev* ('good') and *orthos* ('right').

The other common way of varying the sense of words is suggested by tropes; these include using words as metaphors and similes, synecdochically, metonymically, in opposition (irony) and in conjunction. Taking the notions traditionally found in these two areas, Wilkins reduces them to a table of transcendental particles. In the following table, an example of each transcendental particle is given. These are combined with root words to create derived words:

	Particle plus	*Root word* →	*Derived word*
I. Essential Respects			
1. Similitude	metaphor	root	original
	like	blood	crimson
2. Denoting the nature –	kind	domestic bird	poultry
General	manner	evil	wrong
3. Individual	thing	obscure	mystery
	person	vile	rascal
II. Circumstantial Respects			
1. Position or duration	place	metal	mine
	time	Nativity	Christmas

2. Effecting or representing	cause	to die	kill
	sign (of)	regality	crown
3. Conjunction or separation	aggregate	ship	fleet
	segregate	hour	hourly
III. Of Quantity – Figure			
1. Shape	lamin	wood	plank
	pin	iron	nail
2. Shape and use – More simple	instrument	to dig	spade
	vessel	to drink	cup
3. Less simple	jugament	to weave	loom
	machin	to grind	mill
IV. Of Substance			
1. Enclosure – Places or things	sepiment	sheep	sheepfold
	armament	foot	horseshoe
2. Men – Contiguous to	vest	head	cap
	armour	head	helmet
3. Remote from	house	mad	Bethlehem (i.e. Bedlam)
	room	dunging	privy
V. Of Quality			
1. General quality	habit (of)	to rejoice	cheerfulness
	art	harmony	music
2. Personal – Business faculty, skill	officer	navy	admiral
	artist	star	astronomer
3. Manufacture or exchange	mechanic	stone	mason
	merchant	book	stationer
VI. Of Action or Passion			
1. Ability or disposition	power	to feel	sensitiveness
	aptitude	to break	brittle
2. Beginning or repeating	inceptive	morning	dawning
	frequentive	talk	babble
3. Application of power	endeavour	to hear	listen
	impetus	motion	jerk
VII. Of Relation with respect to Quality and Action			
1. Measure – Great or little	augmentative	to hate	abhor
	diminutive	hot	lukewarm
2. Too much, too little	excessive	plane	mountain
	defective	plane	valley
3. Manner as it should or should not be	perfective	voice	euphony
	curruptive	to write	scribble
VIII. Of Relation with respect to Affections of Animals			
1. Sounds	voice	lion	roaring
	language	Spain	Spanish

2. Sexes	male	ruler	king
	female	dog	bitch
3. Miscellaneous	young	horse	colt
	part	leg	shin

Having Wilkins' scheme before us it is possible once again to consider the difference between Wilkins' and Dalgarno's projects and to consider furthermore the theoretical issues that underlie those differences. As we pointed out, it was Dalgarno's contention that in a philosophical language the radical words should be logically simple notions; complex terms are properly derivatives of these radicals. With the benefit of hindsight, Dalgarno argues that the ideas/things which Wilkins includes in his tables, i.e. the species of things, are complex rather than simple notions. As an example Dalgarno points to a group of words for 'body purging', e.g. vomiting, sweating, bleeding, etc. In a wholly taxonomic system such as Wilkins' these are treated as separate species of a genus, and species or the terminal taxa are considered to be radical, i.e. they are the root words from which derivatives are then composed. What Dalgarno argues is that these words are not in fact radical; they are 'complex notions, each being a combination of a primitive notion of purging or excreting plus the thing excreted'.[45] What is primitive for Dalgarno in taxonomic terms is not the species but the genus. Dalgarno goes on to note that where Wilkins does provide for complex terms, by the combination of a radical and a transcendental particle, he duplicates complex terms so that they are found both among the radicals and among the derivations.[46] There are no clearly defined boundaries between simple and complex ideas and terms, a distinction, Dalgarno argues, that is fundamental if the language is to be logical and/or philosophical.

This is another instance of the problems inherent in the idea of 'simple notions'. What Dalgarno meant by 'philosophical' in a philosophical language was 'logical'; what Wilkins meant by 'philosophical' was 'scientific' or 'biological'. In a biological model, a species is a 'simple notion' – an organic, indivisible entity or essence. The difference is not separable from the species; the species is not a concatenation of features. Even though blueberry and white water lily contain a 'simple notion' in logical or linguistic terms (the genus: berry, water lily), there is no reality to this 'simple notion' of genus in the world of nature. A complex notion in logic is not of the same order as a species in biology. A species in nature is not divisible; the *idea* we have of it, which is what logic treats, is. To use classification as a model of interpreting nature and as a language representing nature necessarily raises these questions about simple notions and about essence, questions that continued to be raised until the end of the century – questions upon which both classification and philosophical languages stand or fall.

ii. Reactions and continuations

Immediately after publication of the *Essay*, Wilkins requested that 'the Royal Society would appoint some of their Number, thoroughly to examine and consider the whole, and to suggest, what they judged fit to be amended'.[47] On 14 May, the Society appointed a committee which included Ward, Boyle, Wallis, Holder, Wren, Merret, Ray, Pope, Haak and Hooke but it is not recorded that the committee ever met.[48]

Outside the Society, the *Essay* was read and discussed in the years immediately following its publication and far into the succeeding years. Wallis was sufficiently familiar with it in 1668 to communicate with Wilkins both in the alphabetic form and in the real character. The words were perfectly understandable, he said, 'as if written in our own language'.[49]

Beck read the *Essay* in the summer of 1668 and found the 'critical and philological part incomparably informed'. He thought the real character would 'be the most compleat that has or can be offered to the world'. But he also expressed fear that 'the capacities of the vulgar will not be able to afford Natural logic or metaphysic enough to manage it...'[50]

John Webb considered Wilkins' language a step toward the restoration of the primitive language and offered qualified praise: 'if others would as willingly contribute their studies, as he hath ingeniously begun...for no humane invention, but Divine creation can make anything perfect on the sudden; we might no longer complain of the unhappy consequences that succeeded the Confusion at Babel'.[51] William Simpson found Wilkins had provided the 'most facile direct road to the improvement of man'.[52]

In 1676 Hooke transcribed a discourse on pocket-watches in the real character to demonstrate its value. Hooke wished that all scientific information could be communicated in this fashion 'it being a Character and Language so truly Philosophical, and so perfectly and thoroughly Methodical, that there seemeth to be nothing wanting to make it have the utmost perfection, and highest Idea of any Character or Language imaginable, as well as for Philosophical as for common and constant use...It being a Character and Language perfectly free from all manner of ambiguity and yet the most copious, expressive and significative of any thing or Notion imaginable, and, which recommends it most to common use, the most easie to be understood and learnt in the World.'[53]

Robert Plot judged Wilkins' project (as well as Dalgarno's) to be 'grounded

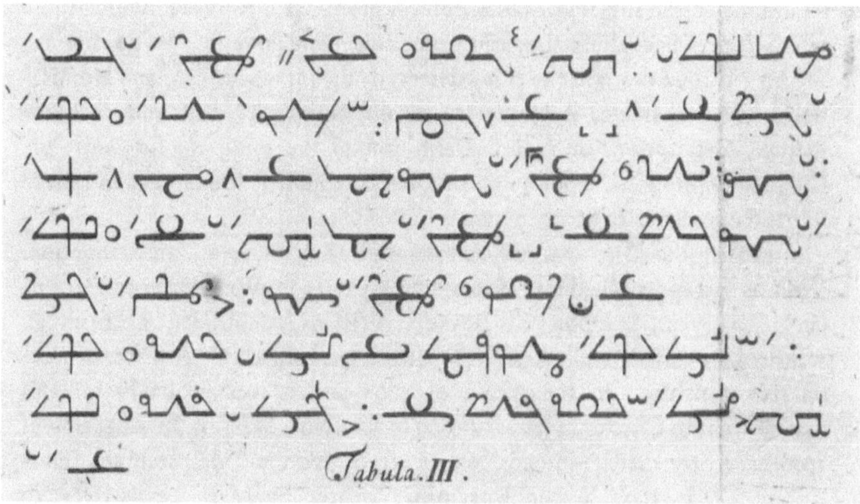

Fig. 8. Hooke's Pocket-Watches[54]

upon rational and solid Principles, with greater Advantages and Facility, than can be believed possible to any that have not made Trial'.[55]

In 1676 Denis Veiras, who had known Wilkins during his association with the Duke of Buckingham, used Wilkins' scheme as the basis for his made-up language in his imaginary voyage, *Histoire des Sévarambes*.[56] It provided the principles for Neremiah Grew's catalogue of the natural history collection of the Royal Society (1681). The preface of the catalogue reveals the attitudes he shared with Wilkins: 'the names of Things should always be taken from something more observedly declarative of their Form, or Nature... For so, every name were a short definition.'[57] The catalogue follows the classification laid out in the philosophical tables of the *Essay*, and descriptions (definitions) are provided in the same manner.

The *Essay* was widely disseminated. Oldenburg provided Comenius with a copy and Wallis sent one to the Italian physicist, Borelli.[58] The Elector Palatine requested one in 1677 and in 1679 the *Essay* is mentioned in Newton's correspondence.[59] Huygens found the design 'fort bien mais tres malaisè a executer, non pas a cause de la difficultè de la chose mais parce qu'il faut le contentement des nations'.[60] Moray replied in turn that the King desired to learn the language and others would follow suit.[61] Locke's friend, Nicholas Toinard, requested a sample of Dalgarno's *Ars signorum* (for a friend 'qui a quelque dessein sur cela') and Locke replied that 'peut etre le livre du Dr. Wilkins...seroit plus a propos...'[62]

Sir William Wotton declared Wilkins' book to be the most considerable treatment on the subject of philosophical grammar: 'A Work, which those who have studied, think they can never commend enough'.[63] The *Athenian Mercury* in 1692 thought the *Essay* deserved 'the highest esteem and Admiration...The Character is Natural, Graceful and Easie, containing a lively Picture, Description, or indeed Definition of the thing it represents; the Language numerous, Copious and Noble, as round as the Greek, and more full and expressive than any extant.'[64]

Interest in the *Essay* was sufficient enough to produce a Latin translation. Wilkins had approached Ray on this as early as 1667. Writing on 13 December of that year, Skippon tells Ray that Wilkins 'is confident no man can translate his book "Real Character" better than yourself'.[65] Ray was at work on this translation in the spring of 1668 and it was probably for this that he went to Chester in January 1669 to stay with Wilkins. In the spring of 1670 he writes that he intends 'to set afresh upon and despatch the translation...'[66] In 1671, a school grammar written by Mark Lewis refers to 'Pictures of Trees, Birds, Beasts and Fishes...in [Wilkins'] Universal Character now coming out in Latin'.[67] Hooke's diary for 29 December 1673 indicates that he 'spoke for Mr. Edw. Pierce, about Latine universal character' and on Christmas Day two years later he 'Looked over Lord Chester's book and Latine translate of Real character'.[68] In September 1676 Andrew Paschall described Ray's Latin tables as almost done.

Ray had finished the Latin version by 1677 for at that time he wrote Aubrey requesting the use of the Latin tables of plants since he planned to publish *Methodus plantarum nova* (1682).[69] Ray's biographer, Derham, reports that in 1716 he found the original manuscript of the translation in the Library of the Royal Society, 'ready for the Presse'.[70] It has subsequently been lost.

A French translation of the *Essay* was planned by Aubrey to be carried out by Lewis Du Moulin, son of the logician, Pierre.[71] There is no indication that the translation was ever made, perhaps owing to Du Moulin's death in 1680.

Despite the mixed reception that greeted the *Essay*, or perhaps because of it, Wilkins planned an amended and enlarged edition. Evelyn understood that Wilkins intended the original *Essay* to be 'preparatory to a more universal work than that of words alone'.[72] Wilkins' appointment as Bishop of Chester at the end of 1668 involved him more in ecclesiastical matters and this may account for the fact that the proposed edition never got it off the ground. In 1670 Ray was persuaded to work on 'altering and amending his tables of natural history'.[73] Evidently he completed these and sent them to Wilkins, but Wilkins died in November 1672 before further work on the enlarged edition was undertaken. Ray reports that after Wilkins' death, no one was found to complete the project.[74] It was a disappointing conclusion since as Evelyn records 'the *Essay* was his Darling and nothing troubled him so much when he dyed, as that he had not completed it'.[75]

In the years following Wilkins' death, a group was formed which was dedicated to revising the real character and philosophical language. What is revealed in their correspondence is the impossibility of taking the project any further. In 1669 Andrew Paschall wrote to Joseph Glanvill, Wilkins' friend and former pupil, that he and some friends had started up a correspondence in the real character.[76] In 1670 Paschall published his *Propositions* for the dissemination of information on the real character and in 1671 he was in touch with Wilkins who apparently approved of the project.[77] Following Wilkins' death Paschall started his own correspondence with Aubrey, a former friend from Cambridge.[78] Around the same time, Lodowyck was working on his 'New Universall Alphabet', an adaptation of Wilkins' phonetic character, and was often discussing it with Hooke.[79] By 1674 Paschall was extolling the virtues of Wilkins' real character for the 'young and untaught'.[80] He tells Aubrey that he has recently seen a child's game based on the universal character, 'a way for making the character standard for other languages' and a way of using it for 'commonplacing', i.e. for storing and retrieving information'.[81] Two years later Paschall sent Aubrey his first revision of one of Wilkins' forty primary genera, suggesting that further revision take the shape of a pocket commonplace book.[82] In 1678 he produced rough drafts of this revision.[83]

This revision consists of five tables: for trees, shrubs and the three classes of herbs. Each table lists the nine genera found in the division in a column down the side of the table. The species for each of the genera are given in a row across, which is partitioned into nine squares. The English names of the species are placed in the squares as well as the real character and the phonetic character for each name. Having drawn up these rough tables, Paschall proposed that they be refined so as to include a picture of the plant and a short description as well as the name of the plant in Greek and Latin. These were to be 'fitted to hang up in the manner of maps in Garden Houses'.[84] The same could be done for the tables of stones, metals, insects, fish, birds and beasts. They could also be put in the form of a pocket book for excursions in nature. Wren approved of the project and Aubrey wrote to Ray about it.[85]

In 1675 and 1676 Hooke had many discussions with Lodowyck about universal language; he wrote his piece on pocket-watches, corresponded with Cave Beck in the real character and 'discussed with Captn. Sheers the universal Character of Navigations and navigators'.[86] In 1676, Aubrey's friend, Thomas Pigott of Oxford, joined in the correspondence and also translated Hooke's pocket-watches.[87] Lodowyck added his part to the discussion of the revisions by pointing out the difficulties he had had in putting terms for trade into the universal character.[88]

The next few years were busy ones for these projectors. In 1677 Paschall had plans that Aubrey in London, Pigott in Oxford, Lodowyck in Holland and an unnamed person in Denmark should publicize and promote the universal character.[89] Hooke told Sir John Hoskins of his 'way of Regulating English to make it a universal language'.[90]

Early in 1678, Hooke talked with Hoskins, Lodowyck and Abraham Hill of a 'catalogue of books and universal language'.[91] With Hoskins, Lodowyck, Dean Lloyd, Mr Smith, and Christopher Wren he 'discoursed much of universal language, his mapps and new shorthand'.[92] In 1678 Petty wrote from Dublin thanking Lodowyck for attempting to put his treatise on duplicate proportion into the real character, which he takes 'to be an honour to Bishop Wilkins and myself, but doubt of its acceptance in the world'.[93]

Aubrey thought of variations on Paschall's revisions which he included in his treatise on education (1678).[94] In addition to Paschall's botanical tables, he would have proverbs set down in real character 'as well as useful and moral questions'.[95] He mentions that Mr Merret of the Inner Temple 'has reduced common law to a method by Dr. Wilkins' way' and that 'The Lord Chief Justice, Hale, hearing of it, desired to see it and was satisfied'.[96]

The following year Hoskins gave 'his notions of universal language' which Hooke characterized as 'very good and most simple'. Mention is also made of Hooke writing a universal language.[97] Pigott was busy in Oxford discussing the universal language with Mr Warden, writing to Dr Lloyd, and corresponding with Lodowyck and Hooke.[98] Edmund Halley joined the circle of those who were 'much persuaded' by the universal character late in 1679.[99]

Although there was a good deal of discussion and endless correspondence there were no projects or proposals that issued from this group of projectors. What happened instead was that the discussion turned to analysis of the principles upon which the philosophical languages of Dalgarno and Wilkins were based, a discussion which reveals further their very different approaches to the project for a universal language.

The discussion was precipitated by Ward who entered into the correspondence of Paschall's group in 1676. Ward criticized the philosophical languages because the letters of the artificial words stood for the numbers of the genera and species rather than for 'transcendental denominations'.[100]

In the next year Ward proposed his own scheme which involved less than a hundred 'transcendentals' or primitive words which were to be compounded to produce the names for the species of things. In February 1677 Pigott recorded his reservations about Ward's proposal.[101]

Pigott argued that while it would be 'most highly philosophical to compound words just as things are compounded' there were too many practical considerations to make this possible.[102] Also, this composition will be arbitrary with each person 'compounding these notions to express his thoughts as he himself speaks' in which case the language will be 'subject to obscurity, equivocation, superfluities, and confused compositions which must needs be as various as the notions and as wild as the fancyes of those men that make them...'[103] Or, even worse, Ward's language scheme will be 'harder than any other [since] we must talk by definitions [and] the speaker and hearer both must resolve a whole discourse into Transcendental principles and run over a whole concatenation of thoughts to express or apprehend one single species of Animals'.[104] He notes that Wilkins had been aware of these problems, having had Dalgarno's example in front of him.

If the composition of these transcendentals is not left to individual fiat, but is formalized in a lexicon, there are problems also: 'I must to the learning of this language be forced to get by heart all these definitions which are but single words perhaps in other tongues.'[105] This, he calculates, is a job which is harder than memorizing the whole Greek lexicon. 'The compositions being longer by so many degrees...and the contexture of the simples exactly to be observed. What though they be agreeable to right reason it is probable I myself would not have drawn them up in the same manner so that I must be obliged to learn your compositions and definitions before I can make them my own...'[106] He concludes that since Wilkins' scheme does not have these problems, additions should be made to Wilkins' tables rather than abandoning his design in favor of Ward's.

The next we hear is that Piggott set out to acquaint himself with botanical classification (probably in connection with Paschall's revisions of the tables of plants) but he reports that he is disinclined to pursue the matter until Ward's ideas are known.[107] In March 1678 Hooke visited Ward and what he learned of Ward's scheme he communicated to Paschall and Pigott.[108] According to Piggott, Ward's scheme inclined 'too much to Lullius whose radices of Bonitas, magnitude, duratio, whose trunchus, Branche, Rami, folia, flores, and fructus can never...make any one reap any fruit from his tree of knowledge'.[109] Paschall went on to read Ritschel's *Contemplationes metaphysicae* as Ward had recommended but this did not help him understand just how

Ward was to make 'a sufficient provision for all things that need names, out of any determinate number of Transcendentals'.[110] Ward's scheme, he felt, was not in harmony with the way in which knowledge proceeds, which a universal language should duplicate: 'As we begin to know a number of the common objects of sense and their modes, and from this foundation or rather centre go on to the accuracies in severall arts and sciences'.[111] Transferring this to a universal language 'we ought regularly give markers or names, first to a competent number of such things and modes and with them to deliver the way and manner of the language as it were for common use and then do the like in the more exact and full accounts that are to be contrived for the several parts into which the universal object of knowledge is to be distributed and divided'.[112] Hooke's letters indicate that he similarly shared Pigott's distaste for Ritschel if not for Lull.[113]

In 1680 in a letter to Hooke, Paschall set out his understanding of the substantial differences between the designs of Wilkins and Ward. Wilkins' purpose, he claimed, was 'to make an orderly enumeration of the several objects of our thoughts and to set each of them in a clear light and then to enable us to give a distinct mark accommodated to the eye and ear, to each of them; and also to direct us how to make use of these in discourse'.[114] Ward's design, on the other hand, is 'to contrast that great number of marks of those objects of our thoughts, by reducing them to generall or transcendental notions (it seems within the number of 100) by which diversely compounded the nature and reason of all those things and notions which our minds are or may be employed about, may be clearly and distinctly represented (not for the use of common discourse but for the service of severe and strict reasoning)'.[115] Paschall compares these differences to 'what is observable in Arithmetic. First we know there is a method of giving markers to every particular number in infinitum 1, 2, 3 etc. where every unit added contributes a new species [i.e., tens, hundreds, thousands, etc.]. And how every species hath a definite name easy for pronunciation... Next in specious arithmetic there are symbols assigned (which may seem to be some of them answering to our Transcendentals) exhibiting so much of this notion of a number as is common to all else.'[116]

After conferring with Ward, Paschall started to develop his own scheme 'for exhibiting the works of God and objects of Cogitation'.[117] He explained to Aubrey that his thoughts had been 'fluctuating between two very different ways. That of my Lord of Sarum [Ward] which is by Transcendentals and another which is by taking a sufficient number of Individuals, one out of each of a sufficient number of heads that are generall...'[118] What he concluded was that 'to the doing of the former, the latter is prerequired...'[119] He argues that he is inclined to 'taking an Individual out of so many kinds as are needful, to give it a name' because that is 'the natural way of our coming to the knowledge of things and imposing names'.[120]

What we know of the scheme is that it consists of twelve categories: Fixed starre and Fire; Sky and Air; Planet and Comet; Meteor; Stone and Metall; Plant; Animal; Man; Angell; Universe; Nothing; God.[121] Each is to be assigned a basic character which 'shal signify so much of the notion or nature of that first head as is in whatever species or individuals that may belong to it, so much as wherein they all agree...'[122] He explains in more detail that there should be

> a reduction of so many things as are thought fit to come under the names of those individuals, with an assignation of the reason why the name is so communicated and some way of distinguishing them all from one another. Thus to the sun are to be reduced all the fixed starres, because agreeing with the sun in substance, action, figure and they may be distinguished by magnitude, situation, distance from fixed points, etc. Thus to the sky about the sun [are to be reduced] the skyes about other fixed starres and to the comet in 1577...To the air about this globe [are to be reduced] the aires about any other planets, as to this globe all other planets, whether those moving about this sun whose number is known, or any others supposed to move about any other sun or fixed starres...[123]

This scheme is clearly different from Ward's in that it partially retains the classificatory schema and Pigott remarked to Aubrey that he 'tremble[s] to think of it coming before Ward'.[124]

Having realized the magnitude of the task before him Paschall relates his reaction to Aubrey: 'I am bewildered in my thoughts about this subject and fallen into rational despair.'[125] He declared that the job of finding out the primitive notions was too great for him, indeed for any man.[126] A year later however he reports to Hooke that he has 'gott over the discourage you mention from the ambiguity of names and signification...I think I have found out so many as that methinks I walke in those writings as in the light, wherein things appear full now and more intelligible.'[127] In August 1682 he repeats to Lodowyck that this scheme of things should be an 'Introduction to knowledge and a repository for reading and observation' so that the character and language should be secondary.[128] In November 1682 he sent Aubrey his scheme but apparently nothing came of it.[129] Shortly thereafter, Aubrey sent Paschall a letter in the real character from Pigott and asked him to translate it for Hooke.[130] This is the last we hear of either Paschall or Pigott.[131]

In Ireland Petty kept up his interest in the universal language. In 1685 at the Dublin Philosophical Society, of which Petty was President, 'a large discourse concerning the formation of letters and an universall character was presented from Mr. J. Keogh...'[132] At the same time, Petty wrote notes for the abortive project 'The Dictionary of Sensible Words' (1685) and in 1686 wrote 'Twelve Theological Words'. As he indicates, 'the Dictionary I have often mentioned was intended to translate all words that are signa rerum et motuum'.[133]

The last work on a universal language is Hooke's 'Mathematical Language', an unpublished fragment believed to have been written around 1686. During the earlier years of the decade, while Hooke was encouraging Paschall's efforts at a revision of the universal language project, he was also discussing the subject frequently, sometimes daily, with Hoskins, Hill, Lodowyck, Pigott, Pope, Peter Perkins, Dr Witty and Dethlev Cluver.[134] As Secretary to the Royal Society he conducted a brief correspondence with Leibniz who had long been interested in the possibility of constructing a universal language that would be analogous to algebra, i.e. that would isolate simple elements and provide rules for their combination. Such a language or character would constitute and could express an analysis of human thought.[135]

Hooke also was concerned with the possibility of improving and developing the universal language in such a way that it would be not only an instrument for representing knowledge, but also an instrument for analysis and discovery as well, 'for the strict and philosophical way of reasoning and inquiry...to regulate, assist and even...compell the mind to find out and comprehend whatsoever is knowable'.[136] This instrument, as he explains, should proceed methodically and go by steps from subject to subject as nature has placed them in order. He had already formulated ideas of this instrument which he called an algebra of algebras, or a science of methods, in his 'General Scheme' of 1666. But as Hooke points out to Leibniz, 'the great difficulty lies in the media', i.e. in determining the proper subjects of inquiry – primitive notions, 'the first alphabet of those which may be cast as symbols are first to be understood'.[137]

Hooke's notions about an algebra of algebras and its relation to a universal language are unfortunately not complete enough to understand fully. The most promising interpretation is that this method was based on a series of natural histories or tables (like Bacon's tables of absences and presences and obviously like Wilkins' and Dalgarno's tables) which ultimately generate an ordered, universal and complete set of hypotheses and provide for their testing.[138] The universal language is then understood as a set of signs which abbreviate these subjects or 'media'. The language, as an instrument of discovery, is predicated on an analysis which determines the simple notions and ranges them in a manner that reveals the ways in which they are related to or lead to one another.

This is a notion that Hooke came back to in 1684 in his treatise 'On Memory'. There he discusses the formation of ideas noting that these dispose themselves in an order such that there is a continued chain of ideas in the brain.[139] This mental ordering of elements presumably matches the ordering of nature; instruments of method and memory, as for example artificial languages, should obviously be formed on their model.

In 'Mathematical Language' Hooke remarks that, in order to construct an artificial language of discovery, names may 'be conveniently ranged according to this order: first the names of all material things in the world...second the names of the parts of such bodies as are most general remarkable; third, the names of remarkable circumstances about the body by which they become sensible to us such as magnitude, position, motion, figure, light, colour, sound, smell, touch, taste etc...'[140] After constructing this 'simple name', Hooke goes on to say that 'we must informe somewhat of what we know concerning it, either as to its being, its acting, or its suffering'.[141] Primitive words, representing primary notions, were to be constructed of two consonants and a vowel. The tables which Hooke appended roughly follow the arrangement of Wilkins' tables, although there are some taxonomic re-arrangements as well as the addition of three new genera (comprising 25 species) of science and arts; trades; instruments, utensils and tools, additions which had been suggested by Leibniz, Lodowyck and Paschall.

As with many of the documents in this last phase of universal language activity, there is not sufficient information to understand or to interpret the significance fully and properly. All we can say is that, with this fragment, attempts further to develop a universal and philosophical language as it was originally conceived come to an end.

This was in 1686, some sixty-five years after Bacon had suggested that a real character might make a useful means of universal communication. In another year Newton would publish the *Principia mathematica* and four years later Locke would come out with the full edition of his *Essay Concerning Human Understanding*. A new stage of science would be under way. By that time, or shortly thereafter, most of the original projectors were dead. Wilkins and Comenius died in the early 1670s. Pigott died in 1686, Petty and Dalgarno in 1687, Ward in 1689, Boyle in 1691 and Lodowyck in 1694. Although Hooke lasted out the century, after the death of his mistress in 1687 he became something of a recluse. The universal language movement had ridden in with the first wave of science; it subsided as the first generation of scientists passed away.

The one who survived and went on to join in the new wave of science was Ray. It is appropriate to end with his comments since as a scientist, taxonomer and colleague of Wilkins, he is admirably suited to speak for the original scientific paradigm that motivated early science and the language projectors.

As it turned out, Ray's taxonomic venture with Wilkins was not altogether a happy one. Right after he completed the tables for the *Essay* he proffered his criticism of Wilkins' project:

> In arranging the tables I was not allowed to follow the lead of nature, but was required to fit the plants to the author's own system. I had to divide herbs into three squadrons or kinds as nearly as possible; I had to divide each squadron into nine lesser kinds of 'differences' as he called them, seeing to it that the plants ordered under each 'difference' did not exceed a certain fixed number; and finally I had to join plants in pairs or otherwise couple them. How could anyone even hope that a method of this sort would be satisfactory and not transparently absurd and imperfect. I frankly say that it was...[142]

A year later he again expressed his dissatisfaction 'especially in being not free to follow nature, but forced to bow and strain things to serve a design according to the exigency of the character'.[143] In 1678, having been told by Aubrey that a French translation of the *Essay* was being planned, Ray was sceptical: 'certainly it would be an excellent work were it not for the two objections I mentioned formerly, the one rendering it hardly practicable; the other rendering the tables lesse naturall and Philosophicall'.[144] What is important to note is that Ray does not criticize the Aristotelian taxonomy on which the project is built; he only criticizes Wilkins' use of it. Should these difficulties be avoided, as Ray was attempting to do in his own work, he believed 'though the character should never be of use in commerce, the tables would certainly be the best systeme or compend of philosophy extant'.[145]

The collaboration of Ray and Wilkins as taxonomers marks the point at which taxonomy as a proper science flourished. Wilkins' tables do not articulate fully or accurately the system of nature, yet they do mark the culmination of a process in which nature and her kinds were removed from the discourse of speech and rearranged in the discourse of the taxonomic table.

Men have always recognized basic groups of things and it is part of the history of a culture that they join these groups together in a hierarchical sequence or arrangement. The deepening of classification beyond the universal, rudimentary categories and the elaboration of taxonomic hierarchy mark developmental stages not only in the history of Western Europe, not only in the history of science, but in the history of culture. These stages appear to be quite regular and predictable. The reasons appear to be regular also – new perceptions, observations and specialized knowledge made possible by the advancement of writing/print/literacy. New ways of knowing give rise to the new kind of representation we call scientific taxonomy.

From the time that Aristotle decontextualized nature and names, a process which found life again in the print culture of the late Renaissance, the movement towards taxonomy had begun. Once nature and her kinds were decontextualized they were bound to be recombined in some orderly configuration. As long as men understood science as natural history and described rather than measured, that configuration would be taxonomy. And until men could reconcile the apparent disparity between organic and inorganic nature, mathematics would be the model for the one, taxonomy the model for the other. What this means is that Wilkins, or at least his endeavors, were necessary, just as were those of Ray. They reassembled the decontextualized things of nature into a coherent discourse. If they had not, someone else would have.

When Ray wrote that Wilkins' tables could be 'the best systeme or compend extant', taxonomy was considered to be a viable scientific procedure equivalent to mathematics and mechanics. Indeed, we see that the taxonomic predispositions of Wilkins and his fellow language projectors were the very same as those of the Royal Society natural historians. It is under these circumstances that Wilkins and his colleagues could believe that the language they were creating was indeed a scientific one. As long as the isomorphic relation between classification and the order of nature remained intact – and it would do so as long as men held an essentialist view of nature – philosophical languages could be thought to be the new language of the real.

Turning from theory to the actual schemes, in the end we are left with no more than odd little words, a fact which is disconcerting. And yet that is where all language ends; the fate of nations can rest on the pinhead of *yea* and *nay*, of two syllables, breaths, really. The fate of the universal language movement rested in the end on little words like *glapa*.

Those little words were to hold up a linguistic edifice that purportedly connected man with the order he so desperately and relentlessly insists on having. Real things in nature give us real, mental concepts upon which real words and real definitions were to be formed. The scheme began with the natural world, the familiar world that exists out there, the same for one and all, and the familiar, recurrent patterns of properties and behavior. By virtue of those recurrences and regularities, man attempts to reduce multiplicity to unity and to bring the objective out-there under the control of his understanding. In that unity he finds the nature and the order of things, an order that establishes, if nothing else, control by his understanding. In a final demonstration of arrogance, delusion, or desperation, have it what you will, he tried to reduce and subjugate language to that order. If he could do so, he could make creation his own. It was a dream as old as man.

All dreams feature certain selected images, and this one was no exception. It took as its image the natural, biological, genealogical order. It is not insignificant that taxonomic concepts and constructs take as their point of departure the most primitive and basic form of order that we have, the order of the family. The word *genus* comes from a word originally related to family, 'what something derives from'; *species* is related to a verb meaning 'to see' or 'spy out'. The genealogical order is something given by the mysterious processes of procreation, of life; it is something confirmed by what we see around us in our first most familiar, familial world. Through the genus we are given our reality, our essence, our nature. Through the external, visible features of mother's eyes and father's chin, (genetic) nature is known and confirmed. The genealogical family is our first (and perhaps our last) intimation of connection, relation and order.

The epistemic structure of taxonomy built upon this image is a tie that binds man to nature. In that bond lie comfort, protection and reassurance. The language which was built upon that episteme was the means by which men tried to subjugate nature to the cozy, known confines of the family circle. As men know their natures by virtue of their genealogy, who they are really and certainly, and as they state it by virtue of their names; so by language created on a genealogical, taxonomic model could they know, name, and state the reality of nature, of the world literally beyond them. It was a happy dream, which, like all dreams had finally to come to an end.

PART III

The end of the taxonomic episteme

CHAPTER 9

The fall of essentialist taxonomy

i. Science, certainty and taxonomy

If we began with the question of what gave rise to the universal language movement, we should no less consider its demise. By the 1690s it was clearly regarded as a relic of the past, as was the essentialist philosophy it was related to. As one commentator put it:

When Bishop Wilkins first undertook this Design, substance and Accidents were a receiv'd Division and accordingly in ranking things, and reducing them to Heads, (which is the great Excellency of this Design) He proceeds according to the Order they stand in, of Substance and Accidents, in the Scale of Predicaments; but were he to begin now, and would suit his Design to the Philosophy in vogue, he must draw a new Scheme, and instead of Accidents must take in Modes, which are very different from Accidents both in Nature and Number. Bishop Wilkins was an extraordinary person, but very projecting; and I doubt this Design may go along with his Daedalus and Archimedes, and be ranked with his Flying Chariot and Voyage to the Moon.[1]

Philosophically, scientifically, socially and otherwise, by the end of the century conditions were not what they had been earlier. By the Revolution of 1688, to take a convenient date, something had happened, something that vastly complicated the optimistic, and simple, world picture of the projectors; something which made the projects for the universal language look like child's play.

For our purposes, the most important developments were in science, where developments in atomistic-mechanistic philosophy challenged the paradigm of Aristotelian physics and ultimately triumphed over it. Since it was Aristotelian first philosophy that conferred legitimacy on taxonomy as a scientific method and endeavor, its demise was the beginning of the end of the taxonomic episteme.

As we have indicated, in its initial stages, English (Baconian) science was predominantly natural history. Its orientation was toward observation and collection of data. As formulated by Bacon, the goal of science was knowledge of the inner essences or forms of things. Such knowledge, he believed, could be obtained with certainty, provided care was taken and the new, empirical method was employed. Certain knowledge of the natures of things achieved through observation and experimentation were the cornerstones of Baconian science. The most natural method that accompanies this scientific outlook is taxonomy, and the implication is that a taxonomic body of scientific knowledge, a knowledge of things and their essences and a set of definitions, can be achieved with certainty. This was the basis for the universal languages.

In ways that were disturbing, however, the atomistic philosophy which started gaining popularity in the 1650s, particularly the theories of Descartes, Gassendi and Hobbes, created problems for empiricism and these notions of certainty. Greatly simplified, the scientific problem in physics reduced itself to the following: 'The mechanical [atomistic] world view purported to explain all natural phenomena by the size, shape and motion of invisible particles of matter yet these particles or corpuscles were not amenable to direct observation, nor could their motions be educed with any mathematical rigour. Physics was, in consequence, necessarily hypothetical...it could never be rigorous.'[2] What these developments in physics meant for science in general was that since the basic stuff of things was imperceptible, there were limits to what could be obtained through observation and sense perception, and there appeared to be limits to the kind or degrees of knowledge which were possible. It was questionable whether certainty and explanation could be achieved.

For the most part, the early Royal Society was firm in asserting its opposition to hypothetical physics and to the dangerous and atheistical implications that could be drawn from the materialistic theories of Descartes and Hobbes. In the course of the second half of the century, various versions of this opposition developed.[3] The 'empirics' rejected all theories and asserted the primacy of natural history – of the minute observation of phenomena and the recording of data. This was particularly applicable to those who studied living things but was not limited to them, and it was the position taken by Locke. Another group, typified by Boyle and Hooke, opted for a middle way. They admitted French atomist theories, but insisted on testing these hypothetical systems through experimentation and observation. It was Boyle's intention to substantiate the atomistic–mechanistic hypothesis to see if it conformed to experience and could be verified by experiment (whether experiment is grounds for certain, demonstrable knowledge is another matter). While, therefore, admitting hypothesis to a science of nature, it is still important to note that for Boyle (and Hooke) observation of things and collection of data were the most fundamental activities of science. They are still firmly within the tradition of natural history.

Prior to Newton's publication of the *Principia*, in the 1670s and 1680s, the issue of certainty was taken up by members of the Royal Society, not only for the obvious reason that the advancement of science required it, but also because the new science – as a rational and orderly explanation of the orderly nature of things – was brought into the service of establishing a model of religious and political organization. The problem with a scientific method that admitted hypotheses was that it compromised the ideal of certainty and hence compromised science. At the same time moreover it was not clear that inductive studies and empirical collections provided certain knowledge either and unless that could be made clear, it was difficult to establish the legitimacy of scientific activity. Thus, specifically in the service of religion, the issue was considered by Wilkins and by his son-in-law, Tillotson, and in science by Wilkins' friend, Boyle, by his pupil, Glanvill, and by his colleague, Locke. Their debate raised the question of the limits of knowledge and the aim of science; that in turn had implications for scientific method and for taxonomy. The end result of this discussion was Locke's *Essay Concerning Human Understanding* which pointed to the contradiction between atomistic philosophy, certain knowledge and taxonomic method.

To the extent that Wilkins was concerned with these issues as scientific issues, he is most closely allied with the 'empirics' or Baconian natural historians. For him, scientific knowledge would come from generalizations raised from particulars and would be based on observation and experiment, not on hypotheses. Hooke implies this when he compares Wilkins' purely taxonomic tables in the *Essay* with his own tables of natural history in the 'General Scheme' which, after performing their function of recording, were meant to raise hypotheses. The business of science for Wilkins was to observe things so as properly to know and understand and define the properties and natures of things.

Wilkins' contribution to the debate over certainty occurs in his treatise, *Of the Principles and Duties of Natural Religion*, a treatise he was composing at the time of his death in 1672 and which was published posthumously by Tillotson in 1675.[4] Wilkins' concern there is to establish a program that steers between the Scylla of scepticism and the Charybdis of dogmatism. While he admits that in principle reality cannot be understood completely by finite understanding, he argues that there are kinds or degrees of certainty which are possible – enough to justify efforts in scientific research and to provide for scientific progress.

Wilkins is concerned, therefore, to set out the limits of what the mind can know. Knowledge or certainty he minimally defines as 'That kind of Assent which doth arise from such plain and clear Evidence, as doth not admit of any reasonable cause of doubting'.[5] There are three degrees of this: absolute infallible certainty, conditional infallible certainty, and indubitable certainty. Absolute infallible certainty, the highest degree, is the prerogative of God. The other two are the kinds attainable by man. Indubitable certainty is the kind of certainty we have in matters of morals. Conditional infallible certainty pertains to science or knowledge of material things. It is based on two conditions: it 'supposes our faculties to be true' and it supposes 'that we do not neglect the exerting of them. And upon such a supposition there is a necessity that some things must be so as we apprehend them and that they cannot possibly be otherwise.'[6] Conditional infallible certainty is subdivided into degrees or kinds: physical certainty and mathematical certainty. Physical certainty is 'that which doth depend upon the evidence of the sense which is the first and highest kind of evidence of which humane nature is capable'.[7] Mathematical certainty 'doth more eminently belong to Mathematical things' but it includes knowledge of other things which 'are capable of the like certainty, namely all such simple abstracted beings, as in their own natures do lie so open and are so obvious to the understanding, that every man's judgement...must necessarily assent to them' as, for example the abstraction that 'the whole is greater than the part'.[8] It is to be noted here that Wilkins places physical certainty higher on the scale than mathematical certainty; this is an evaluation more closely allied to Baconian science and an order which was reversed by many scientists later on.

Physical certainty is based on the evidence of the senses, or on the first faculty of the mind, simple apprehension, 'the perception of such single objects as are proposed to them'.[9] Although Wilkins admits that we do not understand 'the manner of sensation' (this was the point upon which Locke entered the debate) he also argues that this is no reason to deny the fact that it exists: 'Nothing can be more manifest and plain...than that I now see somewhat which hath the appearance of such a colour or figure'.[10] And as we have seen in our discussion of Wilkins' *Essay*, this sense perception is universal, as is also the understanding we have of the objects of that perception:

Now as there is universal agreement in the sensation of outward objects; The Eye and the Ear of all sensitive Creatures, having the same kind of perception of visible and audible things. Those things which appear Green, Blew, or Red to one, having the same appearance to all others. So must it be with the understandings of men likewise, which do agree in the same kind of Perception or simple apprehension of intelligible objects...those kinds of Apprehensions wherein all men do agree, these are called natural Notions.[11]

What we have therefore in Wilkins as a representative of early science is a set of beliefs that dispose him towards taxonomy as a method of science. First, the belief that the highest kind of knowledge we can obtain is observational rather than mathematical. Second, the belief that if treated properly the natural faculties of sense are sufficient for providing the amount of information necessary and hence for providing a satisfactory degree of certainty. Third, the belief that there is a conformity between our perceptions and ideas of things and the natures (essences) of those things. And finally, a belief in essence (held at the same time with a belief in the corpuscular composition of matter) and a belief that the goal of science is the definition of those essences.

Other and later scientists were to see things a bit differently, and in those differences lay the end of the simple optimism about the possibility of understanding nature's workings and the end to the taxonomic episteme. Despite Wilkins' optimism (which was substantial), nature proved to be more recalcitrant than first supposed; theory suggested it might be fundamentally irrational and unknowable. To wrest away its secrets and to validate its order, more powerful theories, instruments and methods were necessary.

The central figures involved in this subsequent change are Boyle, Locke, Ray and Newton. Boyle developed corpuscular philosophy (a kind of empirical atomism) which analyzes observable properties, e.g. color, shape, texture, etc. as configurations of homogeneous, quantitatively differentiated particles. These observable properties are what had been, in Aristotelian philosophy, qualitatively differentiated manifestations of essence on the basis of which genus and species were determined. Boyle's analysis demoted these properties to the status of secondary qualities, accidental and not essential.

Locke developed the implications of corpuscular philosophy: our knowledge must be of nominal, not real, essences. This necessitates a kind of scepticism and excludes the possibility of obtaining certainty or explanation in empirical science; in effect, it discredits taxonomy, which is pre-eminently empirical, as a significant scientific activity.

These Lockean conclusions were applied specifically to botanical taxonomy by John Ray. They pointed to the conclusion that taxonomic systems, and hence taxonomic nomenclatures, were artificial and nominal rather than 'natural' and real. The theories of Locke and the applications of Ray indicated that certain knowledge of nature's order could not be obtained through taxonomy. They showed that there is no necessarily isomorphic relation between words and things, language and nature, nor can there be.

Working independently, but within the atomistic–mechanistic tradition, Newton developed the theories that were put forth in the *Principia mathematica* in 1687. This put an end to the debate between the empirics and the atomists and completed the rejection of Aristotle. Newton's work demonstrated that (scientific) knowledge is gained not through empiricism alone, but through empirical observation in combination with mathematics. Newton formulated the hypothetico-deductive method where demonstration is achieved through mathematical physics; where what is observed is proven necessarily to follow a mathematical–axiomatic system. Newton's work established the pre-eminence of quantitative analysis as opposed to the qualitative analysis upon which classification is based. With Newton's system, the end of science is re-defined, not as the discovery of the nature of things but the prediction of their behavior regardless of whether their natures are known or not. With the rejection of the Aristotelian paradigm of science, classification could no longer be seen as a means of explaining and representing the nature of nature. Taxonomy was supplanted by mathematics as the method and the language of science.

ii. Nominal essences and artificial classification: Boyle, Locke, Ray

Atomism is a kind of physics which stands in direct opposition to the Aristotelian philosophy of nature. It rejects the Aristotelian notion of hylomorphism and consequently rejects the notion of substantial form and essence. In contrast to Aristotelian theory, atomism denies the qualitative differentiation of matter. Instead of matter being composed of the four (qualitatively differentiated) elements, atomists conceived of matter as 'consisting of particles, whose aggregate might be a solid, a liquid an air or vapour'.[12] In atomistic theory, material objects exist independently of non-material, informing elements like soul, form, etc. Differences in substances and their properties and kinds are attributable to the various motions or geometric configurations of these homogeneous particles. Boyle called these particles-in-motion 'corpuscles'; Hooke referred to them as 'little bodies' and 'small machines'. The theory has more recently been described as a world view based on 'little bouncy balls'.

Corpuscular philosophy was a kind of atomism developed in England by Boyle and adopted as the unofficial natural philosophy of the Royal Society.[13] While it was a hypothesis and was clearly at variance with the basic tenets of the empirical Royal Society philosophy of science it was more or less accepted by scientists who questioned it no further and got on with the business of investigating their particular areas of natural phenomena. As explained most fully in *The Origins of Forms and Qualities* (1666) Boyle's corpuscularism may be briefly summarized as follows: 'Matter exists as *prima materia*, "extended, divisible, impenetrable", endowed with motion by God. Matter and motion are at the root of all physical phenomena.'[14] To quote Boyle: 'The phaenomena of nature are caused by the local motion of one part of matter hitting against another...'[15] 'The *prima materia* exists in the form of particles or corpuscles which are divisible mentally or by God but Nature does "scarce ever actually divide" them. The particles form concretions which are actually the primary building blocks of nature. Differences in the size, shape and motion of these constituent concretions account for the differences in colour, odour, taste, etc. of bodies which are found in the visible world.'[16]

What is of interest here is the treatment of the properties of things. We can contrast Boyle's corpuscularism with Aristotelian theory where essence is defined by an attribute that is essential to the existence of the object in question and arises from its very nature or soul. The loss of that essential attribute results in the destruction of that object. In Aristotelian theory, nature is structured in such a way that these essences (and their genera and species) and their attributes persist through change.

For Boyle, however, persistence and fixity of essence is not a law of nature. As he says in *The Origins of Forms and Qualities*, he is 'not fully convinced that there is such a thing as nature's designing to keep such a parcel of matter in such a state that it is clothed with just such essential accidents rather than with any other'.[17] We can take the case of gold, which Boyle argues, is not fixed to retain its color, hardness, etc. What is essential in gold is not these observable properties. What makes gold gold must be the motion and configuration of its imperceptible corpuscles. As Boyle states:

> if, by a bare mechanical change of the internal disposition and structure of a body, a permanent quality, confessed to flow from its substantial form, or inward principle, be abolished, and, perhaps, also immediately succeeded by a new quality mechanically producible; if, I say, this come to pass in a body inanimate, especially, if it be also, as to sense similar, such a phaenomenon will not a little favour that hypothesis, which teaches, that these qualities depend upon certain contextures, and other mechanical affections of the small parts of the bodies, that are endowed with them, and consequently may be abolished when that necessary modification is destroyed.[18]

Accordingly, Boyle distinguishes primary essential qualities (size, shape of the particles and motion) which are imperceptible configurations of particles;

distinct from these are secondary qualities, the other, observable qualities which the primary qualities give rise to.

Translated into Aristotelian terms, this means that all observable properties must be considered accidental; none can be taken as essential. What are 'essential' are the imperceptible configurations and motions. These are the key to the inner workings and real natures of things.

If these ideas are applied to empirical science, a science based on observation, it follows that, when we observe, we do not observe the real, primary properties of things, but only their secondary effects. And for taxonomy it follows that when we classify into kinds we do not classify on the basis of observation of the essential qualities of things. We do not observe likeness or difference in the real (atomistic) nature of things; we see commonly shared surface qualities and we group these together.

While Boyle admits that natural kinds exist in nature (by the provision of God) he concludes that men have not been able to ascertain these real and natural groups; rather they have made, and agreed to, *ad hoc* divisions simply for convenience: 'men having taken notice that certain conspicuous qualities were to be found associated in some bodies, and other conventions of accidents in other bodies, they did for conveniency, and for the more expeditious expression of their conceptions, agree to distinguish them into several sorts, which they call genders or species...as observing many bodies to agree in being fusible, malleable, heavy and the like, they gave to that sort of body the name of metal'.[19]

What we call species therefore are not necessarily groups of individuals sharing in a common essence as the Aristotelians would have it. They are, rather bundles of objects agreeing in observable properties to which a name has been given. The result is that neither taxonomic structure nor the nomenclature built upon it can be taken as representatives of reality:

> I confess, that I have not yet, either in Aristotle or any other writer, met with any genuine and sufficient diagnostic and boundary for the discriminating and limiting the species of things; or to speak more plainly, I have not found that any naturalist has laid down a determinate number and sort of qualities or other attributes, which is sufficient and necessary to constitute all portions of matter endowed with them, distinct kinds of natural bodies: and therefore I observe that most commonly, men look upon these as distinct species of bodies that have had the luck to have distinct names found out for them, though perhaps divers of them differ much less from one another than from other bodies, which (because they have been huddled up under one name) have been looked upon as but one sort of bodies.[20]

Neither the observable properties nor the names guarantee that we are dealing with the real natures (species) of nature.

Where Boyle's corpuscular theories run into obvious difficulty is with living things. To argue that organisms are simply different configurations of homogeneous particles obeying the rules of matter and motion opens one up to the charge of atheism. To argue, however, that these are informed by some kind of soul or life principle is not demonstrable and opens one up to charges of mysticism. On a more practical level to argue that all genera and species are simply words attached to bundles of qualities leaves out the obvious divisions of biological species which have some real basis in the way nature works. Atomistic theory and organisms are not compatible. But it was not Boyle's intention to apply his corpuscular theories to the biological world. Like most of his colleagues, he was content to cite living things as a special case. He argues that God formed organic creatures out of particles before He set the world in motion, and that He determined the specific configurations and motions of particles which we recognize as kinds of real things.[21]

In sum, there are three potential problems in Boyle's corpuscular theory. One is that it is the goal of science to know the natures of things but since those natures consist of configurations of imperceptible particles those natures in principle may not be knowable. The second is that if our notions of genus and species are nominalistic by virtue of our inability to know the real natures of things, then it follows that a taxonomic method must yield results that are nominalistic and artificial rather than real. Finally, an atomistic model is suited to a science of non-living phenomena and in the end requires representation in mathematical language (as Newton was to demonstrate). It is not compatible with a science of living beings which requires a taxonomic model represented in natural language. These ambiguities are found throughout Boyle's works. Thus, while he could 'infinitely deprecate' Aristotelian physics, he could 'much esteem his historical [taxonomic] writings concerning animals'.[22] On the subject of language we have the same thing. Boyle tells us in one place that what is necessary in language is that we 'alter words so that they may better fix the nature of things'.[23] But in another place he could also claim that if the world was 'God's epistle written to mankind' then it was written 'in mathematical letters'.[24]

Philosophically speaking, the most important development in the debate over the nature of scientific knowledge came with the work of Locke who developed the implications of atomistic theory for scientific method, for classification and for a theory of language. The significance of Locke's *Essay Concerning Human Understanding* is that in it he draws together the various lines of thought developed by the scientists: assuming that the nature of nature is atomistic, he explores the question of how the (corpuscular) nature of matter operates on consciousness; how and what it is that we perceive; what the limits of knowledge are given the nature of human 'understanding'; what the proper method of science is given those limits; how on the basis of perception we classify and name. Although not published until twenty years later, Locke's *Essay* originates in the theories and discussions of the Royal Society in the late 1660s and 1670s, particularly in discussions related to corpuscular theory and to the question of certainty in scientific knowledge. We see him following the Royal Society in his adherence to empiricism, his constructive scepticism, his adoption of corpuscular philosophy and his mistrust of language. Where he differed however – and it was a crucial difference – was that he broke the connection between (the ideas in) the mind and the nature of reality. Once that connection was broken, the notion of a language isomorphic with nature, whether that language be a taxonomic structure or the linguistic representations for it – became an impossibility.

Locke's connections with Wilkins, Ward, Boyle, Hooke and other members of the Royal Society go back to the 1650s and 1660s during his days as a student and tutor at Oxford.[25] By the mid-sixties, Locke was closely associated with the experimental group led by Boyle and Hooke. After moving to London in 1667, he was elected to the Royal Society, and thereafter maintained those connections as well as developing a close connection with Wilkins. Both he and Wilkins participated in the Society and both, for example, were members of the committee which arranged experiments.[26]

Other activities tie him to Wilkins as well. He was an intimate friend of Tillotson, Wilkins' son-in-law.[27] More importantly, Locke's patron, Shaftesbury, and Wilkins' patron, Buckingham, were closely associated, being joined by their mutual interests in science as well as in religious toleration. In 1667 Locke wrote an essay on toleration for Shaftesbury who was using him as an advisor in religious and civil matters. An informal committee which included Shaftesbury, Locke, Buckingham and Wilkins was formed to work on the toleration issue. Wilkins was reputed to be the driving force behind it.[28]

We know that Locke's philosophy was informed by the corpuscular theories of Boyle and his ideas on scientific method are consonant with the Royal Society emphasis on empiricism, i.e. collection of data and observation. As far as Locke's linguistic theories go, it is difficult to imagine that, connected as he was at Oxford, Locke did not know of the original discussions on universal language conducted by Ward, Wilkins and Dalgarno.[29] (Both the writings of Boyle and those of Hooke give evidence of their participation in those discussions.) If Locke did not know of proposals for a universal language from his days at Oxford, he surely would have been familiar with Wilkins' *Essay*.[30] And although Locke's *Essay Concerning Human Understanding* mentions no names, there is a rather pointed comment on the idea of an artificial, instituted language; it is difficult not to imagine that this is directed towards the projectors, particularly towards Wilkins.[31]

Around the time that Locke joined the Royal Society, he also came under the professional influence of the physician, Thomas Sydenham. Significantly Sydenham was not a member of the Royal Society, owing no doubt in part to his antipathy for chemical medicine.[32] He convinced Locke of the serious need for a more clinical approach to medicine (as opposed to a theoretical or anatomical orientation). These clinical studies in medicine are analogous to natural history in general science. Locke's interest in formal theories of physics, chemistry and natural philosophy diminished from this point on. His notebooks show him to have been an astute and thorough clinical physician.

In the same way that Aristotle's philosophical notions of classification and his work as a biologist were interdependent, so too were Locke's. As a trained physician, who practiced under one of the greatest medical men of his time, Locke had more than the usual acquaintance with botany and zoology, an interest he kept up throughout his career.[33] It is not coincidence that in explaining his ideas in the *Essay* Locke draws on a wide variety of examples of living things (rather than, say, chemical things, as Boyle had) and the problems of classifying them.

Given his clinical, empirical and biological interests Locke points to a pre-Newtonian theory of science which is taxonomic rather than mathematical. Although it is developed, as were Bacon's and Boyle's, in express reaction against Aristotelian notions of hylomorphism and theories of Aristotelian essence, it was a theory still tied to essentialism (Locke explained real essences as configurations of particles) and to taxonomy. In the philosophy of science, Locke is the transitional figure between Baconian empiricism and Newtonian mathematical physics.

Despite his associations with the Royal Society and many of its members, Locke displayed an ambivalent attitude to that body. He was always connected to it, but at the same time he was tangential to it and somewhat aloof from it. This ambivalence also characterizes his notions of the philosophy of science. His work can be seen to be based on the theories of Boyle and Hooke whose atomism he shared. It reflects also Wilkins' ideas but departs from them in significant ways. His opposition to Wilkins can be found on three points: (1) on Wilkins' optimism about obtaining knowledge through science and, by implication, by classification and observation; (2) on Wilkins' faith in the legitimacy of classification as a scientific method; and (3) on Wilkins' belief that language can be managed so as to be philosophical, i.e. so that it represents the real divisions of nature. Locke understood, in ways which perhaps Wilkins did not, that these were not unrelated.

Locke's *Essay Concerning Human Understanding* was composed over a twenty-year period. Evidence indicates that it dates back to 1671 to his London days when Locke and a group of five or six friends (none of whom were particularly active in the Royal Society) held discussions on the principles of morality and natural religion. Two shorter drafts of the *Essay*, Drafts A and B, date from that year. The title of the drafts, 'Intellectus humanus cum cognitionis certitudine et assensus fermitate' indicates the similarity of the terrain covered by Locke's *Essay* and Wilkins' *Natural Religion.* An epitome of the *Essay* was published in French in 1688 in *Bibliothèque universelle* while Locke was living in Amsterdam. The full English version was published in 1690 following Locke's return to England.[34]

The orientation of Locke's *Essay* is epistemological. The *Essay* raises the question of how we come to know what we (do) know, but in order to answer that question, as Locke tells us in the Preface, he was forced to consider the role played by language or words in the acquisition of knowledge, i.e. how it is that words confuse the issue of knowing, how they cause us to think that we know when we in fact may not. Locke's *Essay*, therefore, involves the triangular relation between the things of nature, our ideas of them and our words for them.

Ontologically, Locke is directly in the tradition of atomistic philosophy as it was developed by his friend Boyle. The following is a representative sample of his acceptance of atomism:

> My present purpose being only to inquire into the knowledge the mind has of things, by those ideas and appearances which God has fitted it to receive from them, and how the mind comes by that knowledge; rather than into their causes or manner of production, I shall not, contrary to the design of this *Essay*, set myself to inquire philosophically into the peculiar constitution of *bodies*, and the configuration of parts, whereby *they* have the power to produce in us the ideas of their sensible qualities. I shall not enter any further into that disquisition; it sufficing to my purpose to observe, that gold or saffron has a power to produce in us the idea of yellow, and snow or milk, the idea of white, which we can only have by our sight; without examining the texture of the parts of those bodies, or the particular figures or motion of the particles which rebound from them, to cause in us that particular sensation: though, when we go beyond the bare ideas in our minds, and would inquire into their causes, we cannot conceive anything else to be in any sensible object, whereby it produces different ideas in us, but the different bulk, figure, number, texture and motion of its insensible parts.[35]

Locke accepts also Boyle's distinction between primary and secondary qualities, i.e. between the qualities that permanently adhere in objects so that the objects cannot be conceived without them (solidity, figure, motion, number) and secondary qualities which are what we perceive by means of the sense faculties (taste, color, smell, etc.). In Locke's version, these qualities are 'powers' inhering in the object, powers to produce certain sensations, powers which 'derive' from the primary qualities. A table, for example, has an objective power to make us see it as brown.[36]

According to Locke's theory of perception, the motions of the particles give off effluences which 'strike the sense organs. This affectation is then carried on to the brain, which in turn affects the mind.'[37] What these 'motions' give rise to in the mind are ideas, i.e. representations, signs, images, etc. of physical objects. There are two types of these ideas present in the mind: simple and complex. By simple idea Locke means: (1) the given, i.e. what the mind receives, as opposed to what it makes; and (2) the indivisible, the atom.[38] Complex ideas are what the mind puts together and this includes ideas for substances.

What is significant for our discussion of classification and language is Locke's discussion of our notions of substances, the particular things of nature. What we know of them are complex ideas, made up from the discrete, simple ideas or simple perceptions we receive through sensation:

> The mind being...furnished with a great number of the simple ideas, conveyed in by the senses as they are found in exterior things, or by reflection on its own operations, takes notice also that a certain number of these simple ideas go constantly together; which being presumed to belong to one thing, and words being suited to common apprehensions, and made use of for quick dispatch, are called, so united in one subject, by one name; which, by inadvertency, we are apt afterward to talk of and consider as one simple idea, which indeed is a complication of many ideas together: because... not imagining how these simple ideas *can* subsist by themselves, we accustom ourselves to suppose some *substratum* wherein they do subsist, and from which they do result, which therefore we call *substance*.[39]

What is important to note here is that our understanding of physical objects is based on our perception of secondary qualities and not on direct observation of the primary qualities or the physical objects themselves. To do this we would need to be able to observe the workings of the particles. These configurations of secondary qualities constitute the *nominal* rather than the *real* essences of things and the ideas we form of objects are of their nominal essences rather than of their real essences.[40]

If the understanding of particular objects is not based on knowledge of real essence, neither is the understanding of kinds or sorts of objects. The ideas we have of general kinds is knowledge of their common forms or nominal essences; it is not knowledge of their real essences:

I would not here be thought to forget, much less to deny, that Nature, in the production of things, makes several of them alike: there is nothing more obvious, especially in the races of animals, and all things propagated by seed. But yet I think we may say, *the sorting of them under names is the workmanship of the understanding, taking occasion, from the similitude it observes amongst them, to make abstract general ideas,* and set them up in the mind, with names annexed to them, as patterns or forms...to which as particular things existing are found to agree, so they come to be of that species, have that denomination, or are put into that *classis*...And what are the essences of those species set out and marked by names, but those abstract ideas in the mind...And therefore the supposed real essences of substances, if different from our abstract ideas, cannot be the essences of the species *we* rank things into.[41]

It is at this point important to note that Locke does not deny the existence of real essences in nature, existing independently of the mind. What he denies rather is our ability to have knowledge of them, i.e. to know their causes, laws and principles.[42]

The distinction between real and nominal essences has obviously applicability to the method of classification. While Locke would not deny that individual things exist in nature, and that there is a natural arrangement of these things, he would deny that men can know these. What men can know are the co-occurrences of qualities; what cannot be known are the inner workings of the particles and the rules that cause their configurations.[43] Once the tie is cut between thought and reality (nature) it is implied that any classifying we do is arbitrary. If our ideas of the properties of things or of the common properties of a group of things are based on nominal essences, they cannot be the basis for a natural, certain, real taxonomy.

The only way around this problem would be if the divisions we make mentally by virtue of our ideas (based on nominal essences) correspond with those real divisions of nature. But because of the defective nature of sense perception and the representational way in which sense data are transformed into ideas, it is only a possibility, not a probability, that such conformity exists.[44]

To examine Locke's ideas on classification it is necessary to look at his notions of particular things, real essences, general ideas (especially species), and the natural arrangement in a hierarchy. It is assumed that particular things have real essences, although the nature of those essences is atomistic. Our general ideas, e.g. ideas of species, are not, however, coextensive with those real essences as is the case in Aristotelian philosophy:

our distinct species are *nothing but distinct complex ideas*, with *distinct names annexed to them*. It is true every substance that exists has its peculiar constitution, whereon depend those sensible qualities and powers we observe in it; but the ranking of things into species (which is nothing but sorting them under several titles) is done by us according to the ideas *we* have of them: which, though sufficient to distinguish them by names, that we may be able to discourse of them when we have them not present before us; yet if we suppose it to be done by their real internal constitutions, and that things existing are distinguished by nature into species, by real essences, according as we distinguish them into species by names, we shall be liable to great mistakes.[45]

The trouble with the Aristotelian tradition, Locke argues, is that it equates species with real essences. For Locke, species and the classifications which are based on them are only convenient fictions:

these boundaries of species are as men, and not as Nature, makes them...It is true that many particular substances are so made by Nature, that they have agreement and likeness one with another, and so afford a foundation of being ranked into sorts. But the sorting of things [being done] by us...I cannot see how it can be properly said, that Nature sets the boundaries of the species of things: or, if it be so, our boundaries of species are not exactly conformable to those in Nature. For we...stay not for a perfect discovery of all those qualities which would *best* show us their most material differences and agreements; but we ourselves divide them, by certain obvious appearances, into species, that we may the easier under general names communicate our thoughts about them.[46]

The examples Locke adduces to demonstrate the questionable nature of our divisions are drawn from biology and the current taxonomical literature:

in all the visible corporeal world we see no chasms or gaps. All quite down from us the descent is by easy steps, and a continued series of things, that in each remove differ very little one from the other. There are fishes that have wings, and are not strangers to the airy region: and there are some birds that are inhabitants of the water, whose blood is cold as fishes, and their flesh so like in taste that the scrupulous are allowed them on fish-days. There are animals so near of kin both to birds and beasts that they are in the middle between both: amphibious animals link the terrestrial and aquatic together; seals live at land and sea, and porpoises have the warm blood and entrails of a hog; not to mention what is confidently reported of mermaids, or sea-men.[47]

In his discussion of species, Locke also criticizes the propagationist criterion for the discernment of true species, i.e. the notion that essential characters are those which are present in plants or animals propagated by seed. For counterexamples, Locke cites instances of hybrids, changelings and monsters which, although reproduced from seed, do not manifest all the characters considered to be permanent and essential.[48]

Locke's criticism of the classification of species can be summed up in the following manner: he argues that our classification of objects is not done in terms of the internal constitutions of things but in terms of their observed properties or nominal essences. As Yolton puts it, 'Whether we are talking about thinghood in general or of particular objects, the internal, causally basic, cohering particles are unknown. The corpuscularian hypothesis enables us to articulate a plausible account of that inner constitution; that account is nevertheless hypothesis, not knowledge.'[49] The same must hold true for a taxonomy.

It would seem clear that on philosophical and on professional grounds Locke was not prepared to accept a taxonomic model as real or explanatory or to be optimistic that with the advancement of learning it could ever become so. He accepted it as practical, necessary, convenient and as one of the givens of mental activity, but that is not the same thing as believing that classification provides an expression or representation of reality. Explanations of the reality of nature were for Locke far more elusive than assumed in this simple optimism of Wilkins and the Aristotelian theories of naturalists like Cesalpino and Ray.

Given that we cannot know the real nature of things, whatever conclusions we draw about natural phenomena are hypothetical and only plausible. While Locke indicates that there is no reason in principle that such knowledge cannot be gained, he rather suspects that it will not.[50] He therefore concludes that natural philosophy as a certain and explanatory body of knowledge will never be possible:

> But whilst we are destitute of senses acute enough to discover the minute particles of bodies, and to give us ideas of their mechanical affections, we must be content to be ignorant of their properties and ways of operation; nor can we be assured about them any further than some few trials we make are able to reach. But whether they will succeed again another time, we cannot be certain. This hinders our certain knowledge of universal truths concerning natural bodies: and our reason carries us herein very little beyond particular matter of fact.
>
> And therefore I am apt to doubt that, how far soever human industry may advance useful and experimental philosophy in physical things, *scientifical* [demonstrative] will still be out of our reach: because we want perfect and adequate ideas of those very bodies which are nearest to us, and most under our command.[51]

What is possible for Locke is natural history, i.e. a collection of instances, of observations and of experiments. Locke avoids scepticism in his realization that with empirical study our knowledge grows. 'The fundamental methodological distinction in Locke is not between the "empirical" and the "rational" ... but between the "historical" (or descriptive) and the "philosophical" (or explicative), that is to say in the 17th century context, between "natural history" and "natural philosophy".'[52]

Locke's theory of language (or more precisely of words and of the names for substances) is based on his first philosophy: the question of the reality of language is the same as the question of the reality of knowledge. According to Locke's doctrine of signs, as ideas are signs of objects, so words are signs of those ideas. As we designate and classify physical objects with our ideas, so it follows that the function of language is designatory and classificatory, once removed. We have indicated that ideas of substances are (complex) ideas of nominal rather than real essences. The same is true of words. They are signs of (our ideas of) nominal essences. There is no tie between language and the real nature of objects. This is the feature that distinguishes Locke's linguistic theory from his predecessors like Bacon and Wilkins.

Locke's treatment of words follows from his desire to inquire into the limits of understanding. This inquiry leads to the realization that a significant amount of error and misunderstanding can be attributed to linguistic sources, as Bacon and Boyle had pointed out. The danger, as Locke saw it, was that when men have words for things it leads them to the belief that the thing exists and that one has knowledge about the thing. Many disputes in science, he feared, went back to unclear signification and he cites a scientific dispute in his own experience that turned on the meaning of the word *liquor*.[53]

Locke begins his discussion of words with the observation that, with the exception of proper names, all words for substances are general. We do not name the particular objects of nature for a number of reasons: we could not retain all of them in our memories; such naming would be useless in communication; such naming does not suit the advancement of knowledge which is necessarily about general things.[54] Our general terms are signs for our general ideas (species) of nominal essences. These are, as it were, ideas of bundles of properties which we understand certain objects to share in common. Our words signify these bundles: 'men making abstract ideas, and settling them in their minds with names annexed to them, do thereby enable themselves to consider things, and discourse of them, as it were in bundles, for the easier and readier improvement and communication of their knowledge, which would advance but slowly were their words and thoughts confined only to particulars'.[55]

Since our general ideas are 'creatures of our own making' it follows that so too are the meanings we impute to words. We decide that a general term shall stand for a certain combination of properties.[56]

Despite the fact that all words signify only nominal rather than real essences, and hence do not signify the real natures of things, words can signify more or less really or precisely. Like his Baconian predecessors, Locke distinguishes between the scholastic and the 'scientific' use of words. The scholastic joins terms in propositions without knowledge of the truth of things; what this produces is verbal or nominal truth. This is often the case also with the common man's use of words: they agree with 'men's ideas of things' but not in conjunction with the 'truth of things'; they are not accurate recordings of observation of objects.[57]

Following the linguistic philosophy of the Royal Society, what Locke favors is a more scientific treatment of language, where the meaning of words is based on what we know about things through observation. For this, 'natural history is to be inquired into'.[58] To aid this process, Locke suggests that a dictionary of natural history be constructed, a proposal that is in keeping with proposals made by many of the language projectors:

> It were therefore to be wished, that men versed in physical inquiries, and acquainted with the several sorts of natural bodies, would set down those simple ideas wherein they observe the individuals of each sort constantly to agree...But a dictionary of this sort containing as it were, a natural history, requires too many hands as well as too much time, cost, pains and sagacity ever to be hoped for; and till that be done, we must content ourselves with such definitions of the names of substances as explain the sense men use them in...But though such a Dictionary as I have above mentioned will require too much time, cost and pains to be hoped for in this age; yet methinks it is not unreasonable to propose, that words standing for things which are known and distinguished by their outward shapes should be expressed by little draughts and prints made of them. A vocabulary made after this fashion would perhaps with more ease, and in less time, teach the true signification of many terms...Naturalists, that treat of plants and animals, have found the benefit this way: and he that has had occasion to consult them will have reason to confess that he has a clearer idea of *apium* or *ibex*, from a little print of that herb or beast, than he could have from a long definition of the names of either of them.[59]

In an educational treatise, Locke follows up these general principles by indicating how they can be put into practice by or for the developing mind. He warns 'those that would conduct their understanding right, not to take any term, howsoever authorized by the language of the schools, to stand for anything till they have an idea of it'.[60] He proposes a school program based on Comenian principles (which he knew well): 'join as much other real knowledge with [the study of language] as you can, beginning with that which lies most obvious to the senses; such as is the knowledge of minerals, plants and animals, and particularly timber and fruit trees, their parts and ways of propagation...But more especially, geography, astronomy and anatomy'.[61]

Although Locke shares the projectors' dissatisfaction with 'the cheat of words', in what must be a direct reference to the language projects, particularly to Wilkins' project, he explicitly rejects the viability of an instituted, universal, philosophical language.

I am not so vain as to think that any one can pretend to attempt the perfect reforming the languages of the world, no not so much as of his own country, without rendering himself ridiculous. To require that men should use their words constantly in the same sense, and for none but determined and uniform ideas, would be to think that all men have the same notions and should talk of nothing but what they have clear and distinct ideas of: which is not to be expected by any one who hath not vanity enough to imagine he can prevail with men to be very knowing or very silent.[62]

Locke's philosophy of science eventually had practical consequences for those studying and systematizing living beings. Interestingly, it was Ray, the taxonomer of Wilkins' philosophical language, who applied Locke's lessons to scientific taxonomy. When Ray drew up the tables of plants for Wilkins, he was a confirmed essentialist. In subsequent years he adopted Boyle's corpuscular philosophy as did most of his scientific colleagues without feeling any discrepancy between atomism and taxonomy. Boyle, it will be remembered, had exempted living things from the implications of corpuscular philosophy and these implications were not made applicable to classification in general or classification of living things in particular. We know, for example, that Ray had read Boyle in 1668 and had not found anything there which disturbed his Aristotelian notions of taxonomy.[63] By 1685 Ray was known to be an 'Epicurean' and a devotee of corpuscularism.[64] As late as 1691, when he published *The Wisdom of God in Creation*, we see Ray invoking corpuscular philosophy but not finding this to be at variance with his belief in the natural taxonomic system of animate nature. Like Boyle, he was able to make a distinction between what applied to the animate and to the inanimate. Quoting Wilkins, he asserts that there is 'a vast difference between Natural and Artificial things'. For the analysis of the latter he admits that he 'incline[s] to the Atomick Hypothesis' and 'attribute[s] the various Species of inanimate Bodies to the diverse Figures of the minute Particles of which they are made up'.[65] But for the former, he holds to the time-honored divisions, i.e. to the notion of qualitative divisions of genus and species upon which taxonomy rests.

From the time of Cesalpino, botanical classification had been given status as a theory because it set out motivated principles for choice of essential characters. Those essential, structural characters were determined by virtue of the essential nature of the plant (having to do with reproduction and nutrition). While arguments had been conducted about which structural parts, e.g. flowers, leaves, roots, etc., were authentically the essential ones, it was generally agreed that non-structural properties, e.g. size, color, location, etc., were non-essential and inconsequential for classification. Locke's theories destroyed the distinction between these two kinds of characters. They were all observable qualities, marks of nominal essence. As such, none of them were essential; they were all equally applicable.

From Cesalpino's time also, the determination of kinds presupposed an Aristotelian, logical method of division, i.e. lower taxa were to be distinguished from higher taxa on the basis of one quality or difference. If applied, however, the theories of Locke suggested that all qualities equally should be considered together in determining kinds. As a consequence, this changes the definition of species from being an expression of substantial form to being a radical plurality of nominal characteristics.

Locke did not apply his ideas to problems involved in biological or botanical taxonomy, although those ideas were applicable. When Ray did however he was forced to abandon the Aristotelian principles upon which the seventeenth-century theory of taxonomy had been built. This also meant abandoning the belief that the real system of nature could be discovered and represented. The changes in Ray's taxonomic ideas are a perfect image of the epistemic distance between Wilkins and Locke.

Ray drew up his first taxonomical tables for Wilkins in 1668. The effect of Locke's ideas did not take hold until the middle of the 1690s. Independently of philosophy, however, during those thirty years, botanical taxonomy experienced empirical problems of the kind identified generally by Locke. What appeared to be simple in theory did not work in practice. Aristotelian taxonomy and the method of division required the selection of motivated, essential characters but Ray discovered that determining or selecting characters which are essential was not altogether as easy as theory would seem to imply.[66]

In his biological work between 1668 and 1694 (when he changed his taxonomic philosophy due to the influences of Locke) Ray developed the Cesalpinian–Aristotelian model of taxonomy.[67] That he accepted the Aristotelian principle that the purpose of classification was the determination of the essences of plants by means of essential characters is evidenced throughout his publications. In 1670 he remarks that 'whoever would attempt to institute method in the sciences, and attempt to distribute particulars into their [proper] places...should have all species enumerated, so that it would be possible to define every single genus by its essential and distinctive characters'.[68] In 1674 Ray discussed the problems involved in determining which characteristics are essential as opposed to accidental, a subject to which he returned in 1682. While it is difficult, he concedes, to arrive at this philosophically, he notes that there is an empirical criterion: essential characteristics are those which are perpetuated through propagation by the seed. Characteristics which turn up in plants propagated by other means (e.g. cuttings) are accidental: 'no surer criterion for determining species has occurred to me than the distinguishing features that perpetuate themselves in propagation from seed. Thus, no matter what variations occur in the individuals or the species, if they spring from the seed of one and the same plant, they are accidental variations and not such as distinguish a species...one species never sprang from the seed of another or vice versa.'[69]

In 1682 Ray published a major theoretical statement of his taxonomic principles, *Methodus plantarum nova*, and in 1686 and 1688 he published volumes one and two respectively of *Historia plantarum*, a classification and description of some 18,000 specimens. The system of *Historia plantarum* follows the one outlined in the *Methodus plantarum nova* and there is evidence to suggest that this system is similar to the one Ray prepared for the Latin version of Wilkins' *Essay*.

In 1684 Ray repeats that his purpose in constructing the system of the *Methodus plantarum nova* was 'by so methodizing [plants] and giving them such certain and obvious characteristic notes of the genera, that it shall not be difficult for any man...to find out infallibly any plant...especially being assisted by the figure of it'.[70] 'In order that an inventory of plants may be begun and a classification of them correctly established, we must begin to discover criteria of some sort for distinguishing what are called "species".'[71] The new method of which Ray writes is an outright statement of theoretical adherence to Cesalpinian principles; this entails strict rejection of ecological and utilitarian criteria. Ray adopts the position that classification should be based on 'similarity and agreement of principal parts' meaning parts based on reproduction and nutrition.[72]

Although Ray pays lip service to Cesalpinian–Aristotelian theory it is important to look at Ray's actual classifications of plants; there we see him deviating from what theory would dictate. The principle that essential characters are perpetuated through reproduction led Ray to assert that, after dividing plants into perfect and imperfect (seed/no seed), perfect plants can be divided into monocotyledons and dicotyledons. These are the number of leaf lobes appearing on the seed plants, which in fact are the plant embryo. As embryos they are closest to the reproductive function of the plant:

the seed, or rather, the embryonic plant contained in this seed, must...be considered the most essential part of the reproductive system because it is the aim of the entire process of generation. This embryonic plant must furnish characters for the first division of plants since it is a distinct individual in which are concentrated all of the differences which will influence its general development and structure. The first discoverable differences in this embryonic plant consists of the number of lobes or cotyledons.[73]

The ideas belong to Linnaeus but they articulate the reasons behind Ray's intuitive division.

Since the cotyledons are the embryos of the plants, they contain within them all the essential characteristics of the plants. But by considering them, Ray implicitly acknowledges that it is the entire complex of plant features which needs to be considered as an expression of the essence. This is, of course, at variance with the requirements of logical division which dictate that one characteristic must be selected.

Ray was not therefore as logical and consistent as Cesalpino would demand; he failed to use systematically a single basis (part related to reproduction) for his taxonomy. Although he showed theoretical loyalty to the logical principles, he also excused his deviation on the ground that, while he could not make a philosophical justification for it, his system as Sloane notes 'seems to preserve empirically perceived groups'.[74]

In the works from his middle period, we see Ray facing a situation akin to fitting a square peg in a round hole: theory demanded that, in order to get a 'natural' method, a true description of reality, one set of principles must be followed; practice showed that following this set of principles resulted in groups that were clearly artificial rather than natural.

Ray continued to labor with the contradiction between the one and the many characters until the 1690s. It was at that time, when he found himself in debates with Rivinus and Tournefort, that Ray seems to have adopted Locke's philosophy.[75] That philosophy, in turn, cut the Gordian knot of contradiction inherent in the Aristotelian taxonomic method. The view that the taxonomer's job was to find the essential characters rested on the assumption that the end of taxonomy was the discovery of real essences. Locke's philosophy provided the justification for rejecting this premise: if there is no form, there are no essential characters. All characteristics are of equal importance and a classification based on multiple characteristics is justified.

Both Rivinus and Tournefort had argued in concert with Cesalpino that in plant classification it is necessary to choose one essential characteristic related to reproduction. In response to the idea that there are no essential characters, Tournefort asserted that 'the author of nature, who has allowed us the liberty of giving whatever names we wish to the genera of plants, has impressed a common character on each of the species, which must serve as a guide for ranging these in their natural places'.[76] Ray, on the other hand, despite what he knew theory demanded, felt that in order to determine the higher classes it was necessary to take into account many characters, i.e. the total aspect and condition of plants: while 'the utility of the characters selected from the flower and fruit [the reproductive parts] is to be conceded for the discrimination of the subalternate genera, for the primary division of plants, I conclude that these are better made by consideration of the total aspect and condition of the plant; or by some other notable property'.[77]

It is said that Ray, in order to establish his position more fully, took up 'philosophical studies', the results of which are to be seen in *De variis plantarum methodis dissertatio brevis*, appended in 1696 to the second edition of *Synopsis methodica stirpium Britannicarum*.[78] It is this tract which shows clear reliance on the Lockean principle that, although essences exist because of the corpuscular nature of matter, we cannot know them. Ray adopts Locke's position that the form in nature and the species of classification are two different things, the latter division having been created by the mind of man for convenient communication:

singular things have been compared among themselves and for the observation of agreements and differences of these, the same name has been imposed on those things agreeing among themselves...all things of which men have ideas are divided into kinds, as it were, groups, called by philosophers *species infimas*. Such division of things into bounded *species infimas*, and the imposition on them of general names is seen necessary for the institution and general use of a common language.[79]

Like Locke, however, Ray is careful to acknowledge that these species have a real existence in nature as do essences: 'although I assert that universals do not exist in the nature of things, but are merely figments of the human mind, I do not deny that they have a foundation in things, particularly in those with agreement or similitude in the principal parts and properties.'[80]

The problem is however that they most basically exist not as substrata in which properties inhere but as configurations of particles formed and set in motion by God: 'species are essentially distinguished from one another and are intransmutable, and the forms or essences of these are either certain specific principles that is, certain very small particles of matter, distinct from all others, and naturally indivisible, or certain specific seminal reasons enclosed by means of an appropriate vehicle'.[81] Because the essences are composed of imperceptible particles Ray follows Locke's claim that they are unknowable:

Since all our knowledge derives from sensation, we know nothing of things which are outside us except through the power that they have to affect our senses in some particular way, and by the mediation of those impressions to cause a particular phantasm to arise in the intellect. If the essences of things are immaterial forms, it is admitted by everyone that these are not encountered in any sensible means. If they are really nothing other than certain fixed proportions and mixtures of elements or natural minima, then these minima can not be perceived by our senses, except with several instruments or aids...Therefore, since the essences of things are unknown to us, certainly the essential generic characters could not be known to us.[82]

On these philosophical grounds, Ray rejects the notion that only reproductive parts should be taken as the essential characters of plants; no characters, in fact, can be taken as essential. They are all equally accidental. Representing the entire organization of the plant therefore seems to Ray compatible with determining the taxa since from the same essences flow the same qualities, operations and other things which are accidents. It is probable that those plants which agree in several attributes conform in their natures.[83] An empirically determined collection of a large number of accidents leads to the determination of taxa even though this violates what had been considered to be proper theory: 'To agree in nature and to agree in genus is the same thing ...for the agreement in nature or genus, there can be no more certain sign or evidence than the possession of several common attributes.'[84]

With the help of Locke's philosophy, Ray is able to provide a theoretical justification for his earlier insight that the characteristics of the species are passed on through propagation; he is able also to justify his major categorization of plants on the basis of the number of cotyledons, i.e. on the characteristics of the embryo which contains the program or essence of the plant. But because of Locke's philosophy Ray is left with the conclusion that his taxonomy and taxonomic method provide not guarantees but only probabilities and approximations of the order of nature.

iii. Universal languages and specialized varieties of language

Not only in science but in whatever area we examine – social organization, economic development, religion, philosophy, politics, education – the last quarter of the seventeenth century in England is significantly different from the age which preceded it. In the late Restoration period, we see a social adjustment that is, on the one hand, a reaction to the egalitarian tendencies of the revolutionary decades, and, on the other hand, the establishment of a pattern of class differentiation and distribution that was to persist until the Industrial Revolution. In education for example, the rate of increase dropped.[85] The reactionary nature of the society was reflected also in the scientific community. In the 1670s and 1680s the atmosphere of the Royal Society was upper class and it became increasingly dilettante. The original practical intentions of the Society diminished in favor of scientific theory.[86] Newton's *Principia* was unintelligible to all but a handful. This specialization in science was counterbalanced by the antagonistic force of neo-classicism in the arts. In the debate between the humanists and scientists, the ancients and moderns, the specialists as it were of the eighteenth century, the world of Snow's two cultures was born.

Changes in attitude about the function of science help to explain the demise of the language projects. One of the main features of Baconian science was that it was to be practical; to be of some use in serving the affairs of men, whether in manufacturing, or in religion, or in language and communication.[87] Thus in the 1640s and 1650s scientific endeavor was seen by Comenius, Hartlib, Lodowyck, Wilkins, Beck and Boyle as serving to spread the word of God as well as to improve man's health, his material welfare, his commercial enterprises, etc. Even the social composition of the early scientific group marks this practical bent: Wilkins, for example, was the son of a goldsmith; Petty's family was in weaving and dying, Boyle's family in ironworking, and Evelyn's in saltpeter. The free mixing of the classes within the scientific group was often (disapprovingly) remarked. Furthermore the scientists at Oxford at first had close connections with the practical world of commercial affairs as represented by Lodowyck and Hartlib. Thus Wilkins could argue in *Mathematicall Magick* that science held 'much real benefit... particularly for such gentlemen as employ their estates in those chargeable adventures of Drayning, Mines, Cole-pits, etc.... and also for such common artificers as are well skilled in the practice of the arts'.[88] Notes for the statutes of the Royal Society firmly declare that its 'business' is 'to improve knowledge of naturall things and all useful Arts, Manufactures, Mechanick practices, Engynes and Inventions by Experiments'.[89] And although the universal languages were associated with the advancement of learning, this was not contradictory to their practical aim of providing a speedy means of religious, scientific and commercial communication.

The official Royal Society of 1660 continued its policy of promoting the virtues of science by promoting its social usefulness, particularly its usefulness for manufacturing and trade, arguing that the improvement was significant for the general welfare.[90] At the same time, however, the Society found itself in the paradoxical position of being in its social composition primarily an aristocratic group. It obtained royal patronage and could list noblemen among its members. Perhaps more significant is the fact that, with the exception of those who were originally connected with the Oxford group, very few members were from the lower end of the middle class, the class engaged in manufacturing and trade. It has been suggested that the reason for this odd situation is based on the politics of survival.[91] Having been associated with the dissenting Puritans during the Interregnum, it was incumbent upon the scientific group led by Wilkins to establish its credibility with the Court and to dispel the suspicion that science was a potentially disruptive if not impious activity. Formally speaking, the membership of the Royal Society bears out this realignment with court circles.

In the period after 1688 the social orientation of the Society changed. Rather than being aristocratic in its allegiance and practical in its purpose, the Royal Society came to be associated with upper middle-class gentility and with amateurism.[91] In general, the social attitude toward scientific learning seemed to be that it could be tolerated as long as it was not taken all that seriously. The difference between the attitude of the first generation scientists like Wilkins and the scientific dabblers who came later is reflected in Locke's comment that a gentleman must 'look into' natural history so as to 'fit himself for conversation'.[92] Scientific activity and knowledge came to be regarded as a diversion and an ornament, much as music or dancing might be. By the end of the century science and its Society did not see itself as serving the ungenteel society at large, particularly the un-genteel society of trade. Furthermore, the *nouveau riche* of the urban bourgeoisie, once admitted into the ranks of the ruling class, turned away from associating with the economic activity that gave it birth. It is obvious that, in this context, the intensely practical and implicitly democratic principles upon which early science was based would find no sympathy. In the later years of the century, these social values died off with the original scientists. The same is true of the practically conceived projects like a universal and philosophical language.

This shift in the attitude to science was also accompanied by shifting attitudes to the specialization of language, whether it be artificial or natural. The period during which the artificial language was conceived and developed was a period in which new developments in science and in technology, in specialized knowledge in general, put tremendous pressure on the resources of the given language. The most pressing necessity was the creation of additional, specialized vocabulary to express new things and notions. While that pressure decreased somewhat as the century wore to an end, the language continued to add terms as science and technology continued to develop. What changed more dramatically than the actual linguistic situation were attitudes towards this newly reformed language. This diversification or specialization which had begun earlier in the century continued to be reflected in language, albeit in different ways. The notion of a universal language had been one response to this specialization; once it was put aside, attention turned to ordinary language. What is found there is the elaboration of (specialized) varieties.

In the earlier years of the century the attitude was one of linguistic accommodation and acceptance. Development of scientific terminology was greeted as a mark of progress just as development of scientific knowledge was. Part of this is attributable to the enthusiasm about establishing the vernacular as the prestige language of the newly developed society. As English was accepted so too was the newer variety of specialized language. In 1664 the Royal Society could officially sponsor a committee on language since there were persons 'whose genius was very proper and inclined to improve the English tongue, particularly for philosophic purposes...'[94] Although it had but short-lived success, perhaps owing to the disruptive effects of the plague in London, it could name among its twenty-two members not only Wilkins but also Dryden, Waller, Evelyn, Sprat and Cowley. Evelyn found no anomaly in associating with Dryden, Waller and Cowley at the same time as considering Hooke, Petty and Wilkins as 'three such persons together [who] were not to be found elsewhere in Europe for parts and ingenuity'.[95] Evelyn himself proposed collections of technical, exotic and dialect words which he felt were 'worthy to be retained in the language'.[96]

The Dryden of the Restoration thought 'the terms of art in every tongue being more of the idiom of it than any other words', it was 'no shame' to learn them and he could praise English as 'noble', having been improved for the better.[97] Sprat could proudly announce that, in stylistic matters, the Society's writers agreed 'to reject all amplifications, digressions and swellings of style; to return back to the primitive purity, and shortness, when men deliver'd so many things, almost in an equal number of words. They have exacted from all their members, a close, naked natural way of speaking; positive expressions; clear senses; a native easiness, bringing all things as near Mathematical plainess, as they can: and preferring the language of Artizans, Countryman, and Merchants, before that, of Wits, or Scholars.'[98]

By the end of the century reaction against English and against specialized or scientific language (technical terms) had set in. Latin regained its status as the language of the educated and the elegant and they did not regard science in any significant way. The reaction against natural history was accompanied by a reaction against specialized language, terms and jargon. Where the earlier age would wish to enrich the capacity of English, Swift would wish, in his characteristic way, to 'purge' it.[99]

The protest about language really boiled down to the place appointed to scientific knowledge in the cultural context. Science was either entirely rejected or it was relegated to a small, odd, insignificant band of professionals whose proper place was behind the closed doors of their laboratories, not about the market place, or the public forum or the salon. The poets banned scientific language from the arts and contrasted to it the language of the socially prestigious, i.e. the Latin-based language of the elegant, cultivated ruling class.

Those who were not as inclined to dismiss science and hence scientific language out of hand nonetheless ascribed it to a limited, specialized social context. Whereas the early Royal Society had hoped to see scientific language provide a model for the language of discourse, Locke assumed that it was relevant only to a small group of experts. The common language, he implied, was sufficient unto the needs of the common man.[100]

To cap it all, it would not be long before scientists themselves realized that their specialized scientific language would not take the shape of ordinary vernacular language. The specialized nomenclatures of botany and zoology were fixed by Linnaeus in an artificial kind of Latin (which is about as close as natural language can get to numbers) while Newton demonstrated that the language of the physical sciences must henceforth be mathematics. Scientific language was rejected for common use and was relegated to a small, esoteric community of experts. Like the artificial language, in no way could it be thought of as 'universal'.

Yet despite these changes, the underlying demand for order did not diminish. It not only persisted but grew and strategies for ordering and controlling proliferate each in its own domain. The picture we have of the later years of the century is one of increased differentiation and specialization, yet these are held in check by the prevalence and persistence of myths and ideas of order. In the social sphere we witness the stabilization of Augustan society after the pendulum swings between revolution and counter-revolution; extremes or imbalances of any sort – be they the powers of the monarchy, or the aristocracy, or the extremes of religious enthusiasm or of dogmatism – are contained, compromised and controlled. In the psyche, order is enforced through the powers of logic, reason and common sense. There is specialization that divides the pursuers of order between Pope's rhymed couplets and Newton's formidable equations.[101] There is specialization in socio-economic distribution that is regulated not by the interdependent parts of an organic body politic, but by the invisible and orderly hand of Adam Smith's market economy. Lovejoy has shown that at no time was the myth of the Great Chain of Being articulated with more frequency and fervor than in the late seventeenth and early eighteenth centuries. Like the lady who protests too much, it betrays the need to exert force and control over the increasingly differentiated order of things.

Language bears witness to this control and is seen as a closed, 'delimitable field'. The idea of order is all pervasive: dictionaries impose order on the burgeoning vocabulary; grammar brings the order of rule to words in sentences. Proposals for an academy attempt to bring language under a centralized, institutionalized authority. The stylistic norms of neo-classicism establish symmetry and balance as the forces of order in the art of discourse. The artificial and philosophical languages represent a similar attempt, one of the first, to reduce the new multiplicity to order. The taxonomic episteme on which they were based was one that identified an existing order of things which was, if not always strictly visible, at least apprehensible and always representable. The universal languages were thus predicted on a world picture which existed in a construct of words, in language, or more appropriately for that recently literate age, in a world set down in a book. They were simply superseded by more extensive, more developed, more powerful and more workable strategies for order.

iv. Conclusion

The period during which the artificial languages arose was a period of immense change – change in man's knowledge of his world and in his social organization. If we have called it an explosion of information it is not merely to be dramatic; it is to try to give a sense of the revolutionary multiplicity with which its participants were confronted. In the face of that explosion came strategies for coping, for bringing the pieces under control. If, in the first half of the seventeenth century, we see the prevalence of an episteme of order it is not by chance that it occurs at that time; the development of a science of order on a scale never before witnessed was simply imperative.

When it comes to imposing order on what we know, there are many different strategies we can adopt; one of the most natural is classification, something the mind is apparently given to do. In the development of seventeenth-century science what we see is the development of classification as a science: the ordering of everyday life raised to the conscious level of theory as a deliberate attempt to deal with the multiplicity of knowledge.

In the early days of science, classification and mathematics developed side by side, neither one making claims over the other; neither one gaining prominence. The Great Chain of Being represented the idea of order for ordinary men; mathematics represented the idea of order to the physical scientists. In between, as a kind of transition or bridge between the ordinary world of knowledge and experience, and the new, inscrutable mechanistic world was natural history and scientific taxonomy, more comprehensive than folklore yet more comprehensible than mathematics.

As a representational system, the universal languages thus rested on this taxonomy which in turn rested on a philosophy of Aristotelian essentialism. Both the philosophy and the method purported to provide a certain, demonstrable analysis of nature, 'beyond reasonable doubt', to use Wilkins' phrase. It was the certainty and hence the universality of this model which made the universal languages possible. And not only possible, but powerful: as a representation of that analysis of nature, they were thought to be representations of reality.

When Ray attacked the theory of fundamental differences, arising from (real) essences and determining (real) species, like Boyle and Locke, he broke the Aristotelian chain from essence to certain knowledge. If what we know are only nominal essences and nominal properties, then the scientific definitions and theories we form of things are, we hope, true but we can not be sure. They are representations of ideas in our minds but not necessarily representations of nature itself. Taxonomy as a representation reveals our divisions, not necessarily nature's divisions. The language that follows from it is similarly a representation of our ideas and not of nature. The isomorphic relation between language and nature, once considered ideal and attainable, is more or less impossible.

And if those real essences (which we cannot know) are composed of homogeneous particles rather than qualitatively differentiated elements, they can be approached not by observation alone but by observation combined with measurement. Classification is not the method by which one can treat these real essences: mathematics is. If there ever will be a language, a set of signs that can provide an image of reality and is isomorphic with it, it must be mathematics and not natural language. The establishment of mathematics as the methodology and symbolic system of science represents the specialization of knowledge and of language carried to its logical conclusion.

By the end of the seventeenth century, mechanistic philosophy and its mathematical method had proved themselves to be the most powerful instrument in scientific explanation. Taxonomy had been demoted as a scientific methodology. It had failed in practice and Locke implied that this failure was necessary given the nature of things. Newtonian physics served the *coup de grâce*. By the end of the seventeenth century, the taxonomer whom Wilkins had so proudly celebrated had become a stock figure of fun, 'employed in the gathering of weeds and the chase of butterflies'.[102] In the end, the break in the isomorphism between taxonomy and nature reduced the universal language projects to metaphors rather than models: rendered them in fact beside the point.

Notes

Introduction

1 Jorge Luis Borges, 'The analytical language of John Wilkins', in *Other Inquisitions*, trans. Ruth L. C. Simms (1964; rpt. New York, 1968), p. 103.

2 The word *character* is used to refer to a writing system; a real character (as in Chinese characters) is an ideograph or ideographic writing system.

3 A. C. Crombie, *Augustine to Galileo* (Harmondsworth, 1969), II, 302–3.

4 Michel Foucault, *The Order of Things: An Archaeology of the Human Sciences* (London, 1970), p. 71ff. Originally published as *Les Mots et les choses* (Paris, 1966).

5 It was not until Bishop Berkeley's time that modern standards were set: 'Neither the Newtonian system nor any other scientific theory could give an account of "the nature of things" or establish the causes of phenomena. Such a physical system was a "mathematical hypothesis"; it established simply the "rules" by which phenomena were found to be connected, and by means of which they could be predicted' (Crombie, *Augustine to Galileo*, II, 329–30).

6 Margaret Jacob, *The Newtonians and the English Revolution 1689–1720* (Ithaca, N.Y., 1976), pp. 16–19.

7 Barbara Shapiro, 'History and natural history in sixteenth- and seventeenth-century England: an essay on the relationship between humanism and science', in *English Scientific Virtuosi in the 16th and 17th Centuries* (Los Angeles, 1979), pp. 23-40.

8 Robert G. Frank, Jr, 'Physician as virtuoso in seventeenth century England', in *English Scientific Virtuosi*, pp. 92–101.

9 Arthur O. Lovejoy, *The Great Chain of Being* (1936; rpt. Cambridge, Ma., 1973), esp. pp. 227–41.

10 Foucault, *Order*, pp. 50ff. and pp. 128ff.

11 Jack Goody, *The Domestication of the Savage Mind* (Cambridge, 1977), esp. pp. 52–111.

12 Lawrence Stone, 'Literacy and education in England 1640–1900', *Past and Present*, 42 (1969), 69–139 and 'The educational revolution in England 1560–1640', *Past and Present*, 28 (1964), 41–80; David Cressy, *Literacy and the Social Order* (Cambridge, 1980), esp. chapter 8.

13 Foucault, *Order*, pp. 157ff.

1. The Aristotelian origins

1 R. G. Collingwood, *The Idea of Nature* (1945; rpt. Oxford, 1965), pp. 97–8.
2 A. C. Crombie, *Augustine to Galileo* (Harmondsworth, 1969), I, 85.
3 *Ibid.*
4 *Ibid.*, p. 87.
5 Aristotle, *Categoriae and De interpretatione*, ed. and trans. E. M. Edghill (Oxford, 1963) in *The Works of Aristotle*, ed. W. D. Ross, 12 vols (London, 1908–52), vol. I. On the categories, see Sister Marina Scheu, *Categories of Being in Aristotle and St. Thomas* (Washington, D.C., 1944), pp. 5–7; John Cook Wilson, 'Categories in Aristotle and in Kant', in *Aristotle: A Collection of Critical Essays*, ed. J. M. E. Moravcsik (Garden City, N.Y., 1967), pp. 75–89; Moravcsik, 'Aristotle's theory of categories', in *Aristotle*, ed. Moravcsik, pp. 125–45; Gilbert Ryle, 'Categories', in *Logic and Language*, ed. Anthony Flew (Oxford, 1953), 2nd series, pp. 65–81; J. Owens, 'Aristotle on categories', *Review of Metaphysics*, 14 (1960–1), pp. 73–90; William Kneale and Martha Kneale, *The Development of Logic* (Oxford, 1962), pp. 25–32. In *Ramus, Method, and the Decay of Dialogue* (Cambridge, Ma., 1958) Walter Ong reminds us that the idea of categories derives from 'accusations' that can be brought against a subject, that is to say, that they were originally auditory. The later tradition of Agricolan place logic (where the categories were assimilated to 'places') betokens a shift in emphasis to the visual which is associated with written language rather than with speech.
6 Jacques Maritain, *An Introduction to Philosophy*, trans. E. I. Watkin (London, 1930), p. 191.
7 *Ibid.*, p. 200.
8 David Pears, 'Grasping the Essentials', *Times Literary Supplement*, 13 February 1976, p. 174.
9 Maritain, *Introduction*, pp. 221–2n.
10 Marjorie Grene, *A Portrait of Aristotle* (Chicago, 1963), pp. 73–4.
11 W. D. Ross, *Aristotle* (1923; rpt. London, 1960), p. 23.
12 For the original see *Porphyrii isagoge et in Aristotelis categorias commentarium*, ed. A. Busse (Berlin, 1887) in *Commentaria in Aristotelem Graeca*, Academiae Litterarum Regiae Borussicae (Berlin, 1887), IV(i). I have used Porphyry, *The Organon or Logical Treatises of Aristotle with the Introduction of Porphyry*, ed. and trans. Octavius F. Owen (London, 1853), II, 611. See Aristotle, *Topica*, 102a 31.
13 Porphyry, *The Organon*, ed. Owen, I, 613.
14 *Ibid.*, pp. 618–19.
15 Grene, *Portrait*, p. 87.
16 *Porphyrii isagoge*, ed. Busse, ch. 1, ll. 8–13. Translation from Porphyry. *The Organon*, ed. Owen, II, 610.
17 'Universals', in *The Encyclopedia of Philosophy*, ed. Paul Edwards (New York, 1967), VIII, 203.
18 *On Interpretation*, trans. Harold Cook (Cambridge, Ma.: Loeb Classical Library, 1962), 16a 1–9. Language in this context means words and words mean names. Syncategoremic terms, prepositions, conjunctions and some adverbs are excluded.

19 Thomas Hobbes, *Leviathan* in *The English Works of Thomas Hobbes*, ed. William Molesworth (London, 1839–45), VIII, 21.
20 Hobbes, *Elements of Philosophy* in *The English Works*, I, 19–20.
21 *Ibid.*, p. 17.
22 *On Interpretation*, trans. Cook, 16a 13–15.
23 *Ibid.*, 16a 16–17.
24 Thomas Wilson, *The Rule of Reason Conteinyng the Arte of Logique* (London, 1551), D7v. There is a modern facsimile published by de Capo Press (New York, 1970) and a critical edition edited by Richard S. Sprague (Northridge, Ca., 1972).
25 John Robert Cassidy, 'Aristotle on definitions', *Southern Journal of Philosophy*, V (1967, p. 115. See also Karl R. Popper, *The Open Society*, 4th edn (revised), (Princeton, N.J., 1963), I, 31–4; II, 9–21.
26 *Analytica posteriora*, ed. and trans. G. R. G. Mure (Oxford, 1963) in Aristotle, *Works*, I, 93b 30.
27 *Ibid.*, 92b 26–32.
28 *Metaphysica*, ed. and trans. W. D. Ross (Oxford, 1963) in Aristotle, *Works*, VIII, 1030a.
29 *Ibid.*, 1045a 12.
30 *Posterior Analytics*, trans. Hugh Tredennick (Cambridge, Ma.: Loeb Classical Library, 1961), 93b 38–9.
31 *Ibid.*, 94a 5.
32 Crombie, *Augustine to Galileo*, I, 82.
33 Collingwood, *Idea of Nature*, p. 98.
34 Alfred North Whitehead, *Science and the Modern World* (1925; rpt. New York, 1967), p. 21.
35 *Ibid.*, pp. 19–21.
36 *Ibid.*, p. 28.
37 *Ibid.*
38 *Ibid.*
39 D. M. Balme, 'Genos and eidos in Aristotle's biology', *Classical Quarterly*, 12 n.s. (1962), 81–98.
40 Porphyry, *The Organon*, ed. Owen, II, 614–16.
41 *Metaphysica*, 1038a.
42 *Ibid.*
43 *Ibid.* It should be noted that in this kind of division one seeks to determine the constitutive species of the genus, and each species is 'defined' by some essential difference. This is why Aristotle objected to Plato's method of dichotomous division, where each constitutive difference is of the form X and not -X (e.g. winged, wingless). A species must be determined by a fundamental difference and 'wingless' or 'non-animal' or 'non-mortal' cannot be essential differences. Furthermore in Plato's definition of man (animal, mortal, footed, biped, wingless) the character 'wingless' does not proceed from the previous difference.
44 Porphyry, *The Organon*, ed. Owen, II, 614: 'Substance indeed, is itself genus, under this body, under body animated body, under which is animal, under animal rational animal, under which is man, under man, Socrates, Plato and men particularly.' See I, 7 for a drawing of the Porphyrian tree.

45 Thomas Blundeville, *The Art of Logicke* (London, 1599), p. 19.
46 Pierre Du Moulin, *The Elements of Logick written first in Latine, then in French... and now translated into English* by Joshua Ahier (Oxford, 1647). Du Moulin's sons, Lewis and Peter, were colleagues of Wilkins at Oxford.
47 Ross, *Aristotle*, p. 115. See *Historia animalium*, ed. and trans. D'Arcy Wentworth Thompson (Oxford, 1912) in Aristotle, *Works*, IV, 486a 14–b22, 497b 6–13; *De partibus animalium*, ed. and trans. William Ogle (Oxford, 1912) in Aristotle, *Works*, V, 644a 16–23.
48 Ross, *Aristotle*, p. 115.
49 *Ibid.*
50 Crombie, *Augustine to Galileo*, I, 165.
51 *Ibid.*, p. 149.
52 *Ibid.*
53 *Ibid.* See *Historia animalium*, 588b 4.
54 Ross, *Aristotle*, p. 116. See *De generatione animalium*, 732a 25–733b 16.
55 Ross, *Aristotle*, pp. 116–17.
56 *Ibid.*, p. 117.
57 Theophrastus, *Enquiry into Plants* (*Historia plantarum*), trans. Sir Arthur Hort (London, 1916), Bk I.
58 *Historia animalium* 588b 5–17.
59 *De partibus animalium*, 681a 10–15.
60 'I call two things continuous when the limits of each, with which they touch and by which they are kept together, become one and the same, so that plainly the continuous is found in the things out of which a unity naturally arises in virtue of their contact' (*Metaphysica*, 1069a 5). 'Instances of discrete quantities are number and speech; of continuous, lines, surfaces, solids, and, besides these, time and place' (*Categoriae*, 4b 22).
61 Lovejoy, *Great Chain*, p. 59.
62 *Ibid.*
63 *Ibid.*
64 Thomas Sprat, *The History of the Royal Society* (London, 1667), p. 110. There is a facsimile edited by Jackson I. Cope and Harold Whitmore Jones (St Louis, Mo., 1958).

2. Taxonomy

1 See Marjorie Grene, 'Aristotle and modern biology', in *Topics in the Philosophy of Biology*, ed. Grene and E. Mendelsohn, Boston Studies in the Philosophy of Science (Boston, 1976), 27, 411–12: 'the role in biology of a concept of form, organization or information should demonstrate...the irreducibility of biology to physics and chemistry...This was Aristotle's thesis also, against Democritus. Organized systems cannot be understood in terms of their least parts alone, but only in terms of those parts *as organized* in such systems. Organized systems are *doubly determinate;* they exist on at least two levels at once...There may or may not be a cosmic hierarchy; the fact remains that whenever we study living systems we are studying particular, limited systems that are hierarchically organized, organized on at least two levels. We are studying systems composed of elements

obeying their own laws, but constrained at the same time by arrangements of those very elements which constitute, as such, laws of a higher level. The higher level – form, organizing principle, code, fixed action pattern or what you will – exists only *in* its elements, and depends on them for its continuance, yet the laws of the elements in themselves...permitting any number of informing arrangements, do not as such account for the principle which in this case happens to constrain them.'

2 R. K. Merton, *Science, Technology and Society in 17th Century England* (1938; rpt. New York, 1970), pp. 45–53.

3 Goody, *The Domestication of the Savage Mind* (Cambridge, 1977), p. 6.

4 *Ibid.*, pp. 37–73. On the relationship of literacy to philosophy, logic and science see also Eric Havelock's seminal study, *Preface to Plato* (Cambridge, Ma., 1963), esp. pp. 215–33 and pp. 295ff. See also Havelock, *Origins of Western Literacy* (Toronto, 1976), esp. pp. 78ff. and Walter J. Ong, *Interfaces of the Word: Studies in the Evolution of Consciousness and Culture* (Ithaca, N.Y., 1977), esp. pp. 17ff. and 121ff.

5 Goody, *Domestication*, p. 51.

6 Eric Havelock, *Prologue to Greek Literacy* (Norman, Okla., 1973), pp. 378ff.; Havelock, *Origins*, pp. 178–89; Goody, *Domestication*, pp. 36–51.

7 Goody, *Domestication*, p. 78.

8 Havelock, *Preface*, passim.

9 Frances Yates, *The Art of Memory* (Harmondsworth 1969).

10 Ong, *Ramus, Method, and the Decay of Dialogue* (Cambridge, Ma., 1958); 'From allegory to diagram in the Renaissance Mind', *Journal of Aesthetics and Art Criticism*', XVII (1959), pp. 423–40; 'Tudor writings on rhetoric, poetic and literary theory', in *Rhetoric, Romance and Technology: Studies in the Interaction of Expression and Culture* (Ithaca, N.Y., 1971), pp. 48–103; 'System, space and intellect in Renaissance symbolism', *Bibliothèque d'Humanisme et Renaissance*, 18 (1956), pp. 222–39.

11 C. S. Lewis, *The Discarded Image: An Introduction to Medieval and Renaissance Literature* (Cambridge, 1964), p. 10.

12 Stone, 'Literacy and education in England 1640–1900', *Past and Present*, 42 (1969), p. 125. See also Stone, 'The educational revolution in England 1560–1640', *Past and Present* 28 (1964). The reasons for the rise in literacy are many. Among them Stone cites: 'Humanist theories of both elite and mass education, the enormous impact of the printing press, the desire of Puritans for access to the Bible, a new demand for an educated nobility for state service and an educated gentry for the professions, and finally, a pattern of generous educational charity headed by well-to-do childless merchants...' ('Literacy', p. 136). There is a growing body of work on the history of literacy. David Cressy's *Literacy and the Social Order* (Cambridge, 1980) is the most recent and contains the relevant references. Cressy's study came to my attention too late to be included here.

13 Adapted from Stone, 'Literacy', p. 121 and Roger Schofield, 'Dimensions of illiteracy, 1750–1800', *Explorations in Economic History*, 10 (1973), p. 445.

14 Ong, *Ramus*, p. 164.

15 Sister Patricia Reif, 'The textbook tradition in natural philosophy, 1600–1650', *Journal of the History of Ideas*, 30 (1969), p. 22.

16 Murray Cohen, *Sensible Words: Linguistic Practice in England 1640–1785* (Baltimore, Md, 1977), pp. 5ff.

17 Robert Robinson, *The Art of Pronunciation* (1617); John Wallis, *Grammatica linguae Anglicanae. Cui praefigitur de loquela sive sonorum formatione, tractatus grammatico-physicus* (Oxford, 1653); William Holder, *Elements of Speech* (London, 1669); Lodowyck, 'An Essay Towards an Universal Alphabet' (1686); Wilkins, *Essay Towards a Real Character and a Philosophical Language* (1668) pp. 357ff. For orthography: Edmund Coote, *The English Schoolemaister* (1596); Joseph Prat, *The Order of Orthographie* (1662); Joseph Brookbank, *An English Monosyllabary* (1651); Thomas Hunt, *Libellus orthographicus* (1661); Owen Price, *English Orthographie* (1668). For a complete listing see R. C. A. Alston, *The Bibliography of the English Language from the Invention of Printing to the Year 1800* (Leeds, 1965–1973), vol. IV.

18 G. A. Padley, *Grammatical Theory in Western Europe 1500–1700: The Latin Tradition* (Cambridge, 1976).

19 Karl R. Popper, *Conjectures and Refutations: The Growth of Scientific Knowledge* (1963; rpt. New York, 1968), pp. 103, 106. According to Popper the essentialist interpretation of Newton belongs to Roger Cotes in his preface to the 2nd edition of the *Principia*.

20 Reif, 'Textbook tradition', p. 23.

21 *Ibid.*, p. 26.

22 Michel Foucault, *The Order of Things: An Archaeology of the Human Sciences* (London, 1970), p. 136.

23 *Ibid.*, pp. 132–3.

24 Mark Lewis, *An Essay to Facilitate the Education of Youth, by Bringing Down the Rudiments of Grammar to the Sense of Seeing* (London, 1670?).

25 Murray Cohen, *Sensible Words: Linguistic Practice in England 1640–1785* (Baltimore, Md, 1977), pp. 13ff.

26 For examples, see Ong, *Ramus*, passim.

27 Care must be taken to understand that these are not constituents in a mathematical sense, where, for example, the constituents of six are two and four. There was some difficulty in understanding the difference between these two among the language projectors – a difficulty which reveals itself in the various discussions of the simple notions which the philosophical languages were to represent. The same term 'simple notions' is used by Descartes, Dalgarno, Mersenne, Wilkins, Ward and Kinner. But it would appear that, for some, simple notions meant constituents like mathematical or atomistic ones; for others it meant generic constituents. We will take this up later in the text.

28 Dioscorides, *De materia medica; libri quinque*, ed. Max Wellmann (Berlin, 1958); *The Greek Herbal of Dioscorides; Illustrated by a Byzantine A.S. 512, Englished by John Goodyer A.S. 1655*, ed. R. T. Gunther (1933; rpt. Oxford, 1934).

29 A. R. Hall, *The Scientific Revolution* (1954; rpt. London, 1967), p. 277.

30 *Ibid.*, p. 278.

31 Crombie, *Augustine to Galileo* (Harmondsworth, 1969), I, 157.

32 Bartholomaeus Anglicus, *On the Properties of Things: John Trevisa's translation of Bartholomaeus Anglicus de proprietatibus rerum*, ed. M. C. Seymour, et al. (Oxford, 1975–), I, II, passim.

33 Frank J. Anderson, *An Illustrated History of the Herbals* (New York, 1977), p. 111.

34 Ulisse Aldrovandi, *Ornithologiae, hoc est de avibus historiae libri XII* (Bologna, 1660), II; *Aldrovandi on Chickens: The Ornithology of Ulisse Aldrovandi (1600), Vol. II, Bk XIV*, trans. L. R. Lind (Norman, Okla., 1963).

35 Elizabeth Eisenstein, *The Printing Press as an Agent of Change: Communications and Cultural Transformations in Early Modern Europe* (Cambridge, 1979), I, 79.

36 Albertus Magnus, *The Book of Secrets of Albertus Magnus of the Virtues of Herbs, Stones and Certain Beasts, also a Book of the Marvels of the World*, ed. Michael R. Best and Frank H. Brightman (Oxford, 1973), p. 31.

37 Agnes Arber, *Herbals: Their Origin and Evolution* (Cambridge, 1938), p. 171.

38 Banckes, *An Herbal 1525 ed. and transcribed into Modern English with an Introduction*, Sanford V. Larkey and Thomas Pyles (New York, 1941), pp. 8–9. Based on the original edition.

39 John Gerard, *The Herbal or General Historie of Plants* (London, 1597), p. 108.

40 Arber, *Herbals*, pp. 250–1.

41 Anderson, *Illustrated History*, p. 197.

42 Arber, *Herbals*, p. 252.

43 William Coles, *Adam in Eden or Nature's Paradise: The History of Plants, Fruits, Herbs and Flowers* (London, 1657), p. 100.

44 Caspar Bauhin, *Prodomus theatri Botanici* (Frankfort, 1620) in Crombie, *Augustine to Galileo*, II, 269–70.

45 Carolus Linnaeus, *Genera plantarum*, 5th edn (1754), ed. W. T. Stearn (Weinheim, 1960), p. 453. This is a facsimile edition.

46 William Ivins, *Prints and Visual Communication* (Cambridge, Ma., 1953), pp. 43–4.

47 *Ibid.*, p. 44.

48 Brent Berlin, Dennis Breedlove, Peter H. Raven, 'General principles of classification and nomenclature in folk biology', *American Anthropologist*, 75 (1973), pp. 124–216.

49 Arber, *Herbals*, p. 164.

50 *Theophrasti historia plantarum liber primus*, trans. Theodore Gaza (Travisii, 1483).

51 Dodoens published his *Crŭÿdeboeck* in Antwerp in 1554. This was translated into English by Henry Lyte in 1578 and the 1619 edition of Lyte's translation was entitled *A Niewe Herbal*. The *Crŭÿdeboeck* was the basis for Dodoen's *Stirpium historiae pemptades sex* (Antwerp, 1583) in which this classification appeared. Dodoen's herbal also became the basis for Gerard's *Herbal* (1597) in which we find the following classification: I. Grasses, Rushes, Corn, Flags, Bulbous or Onion-rooted Plants; II. Herbs for meat, medicine or sweet smelling use: III. Trees, Shrubs, Bushes, Fruit-bearing Plants, Resins, Gums, Roses, Heath Mosses, Mushrooms...Coral.

52 For a convenient outline of Bauhin's taxonomy see S. H. Vines, 'Robert Morison and John Ray', in *Makers of British Botany*, ed. F. W. Oliver (Cambridge, 1913), pp. 14–15.

53 See Julian von Sachs, *The History of Botany 1530–1860*, trans. Henry Garnseey (Oxford, 1906), p. 35: 'he shares with his predecessors the peculiarity of not distinguishing the larger groups...by special names or by descriptions; it is only from the order in which the species follow one another that we can gather his views on natural relationships. It follows therefore that the natural families, so

far as they are distinguishable in Bauhin's work, have no sharp bounding lines; we might almost conclude he purposely avoided assigning such limits, that he might be able to pass without interruption from one chain of relationship to another.'

54 Frank Anderson, *Illustrated History*, p. 230.

55 John Parkinson, *Theatrum botanicum: The Theater* [sic] *of Plants or An Herball of a Large Extent* (London, 1640), p. 2.

56 James Larson, *Reason and Experience* (Berkeley, Ca., 1971), p. 42. See Andrea Cesalpino, *Questionum Peripateticorum*, in Maurice Dorolle, *Questions péripatéticiennes* (Paris, 1920), p. 106.

57 Cesalpino, *De plantis libri XVI* (Florence, 1583), in Dorolle, *Questions*, p. 74.

58 Cesalpino, *De plantis*, p. 26: 'Cum autem formarum similitudines, & dissimilitudines queramus, ex quibus constat plantarum substantia...' See Sachs, *History*, p. 52.

59 *Ibid.*, p. 1: 'Quoniam autem altricis animae opus est gignere quale ipsum, sive id ex alimento fiat ad conservationem singulorum, sive ex semine ad specierum aeternitatem: duae ad summum partes perfectioribus datae sunt, maxime necessariae: Una, per quam sumant alimentum, quae radix apellatur. Altera qua fructum tanquam fetum ad speciem propagandam ferant...' See Sachs, *History*, pp. 43–4.

60 Sachs, *History*, p. 55.

61 *De plantis*, p. 29: 'Quaecunque autem neque ad totius plantae, neque ad fructus constitutionem faciunt, ut colores, odores, sapores, & alia huiusmodi, accidentia sunt...' See Sachs, *History*, p. 55.

62 For an explanation and tabulation of Cesalpino's arrangement see Vines, 'Robert Morrison', pp. 12–13. See also Dorolle's introduction to *Questions péripatéticiennes*, pp. 81ff.

63 Joachim Jung, *De plantis doxoscopiae physicae minores*, ed. Martin Fogel (Hamburg, 1662) and *Isagoge phytoscopia* ed. Johann Vagetius (Hamburg, 1678). For Jung's life see *Biographie universelle*, 2nd edn, ed. M. M. Michaud (Paris, 1880), XXI, 330–1; Yates, *The Rosicrucian Enlightment* (London, 1972), pp. 91, 126–7; *Dictionary of Scientific Biography*, ed. C. C. Gillispie (New York, 1970–80), VII, 193–6; *Encyclopedia of Philosophy*, ed. Paul Edwards (New York, 1967), IV, 298; G. H. Turnbull, *Hartlib, Dury and Comenius* (Liverpool, 1947), passim; Hans Kangro, *Joachim Jungius Experimente und Gedanken zur Begründung der Chemie als Wissenschaft* (Wiesbaden, 1968), pp. 336–42.

64 Thomas Sprague, *The Evolution of Botanical Taxonomy from Theophrastus to Linnaeus*, Linnean Society (London, 1950), pp. 1–23. The *Isagoge* was reproduced in Ray's *Catalogus plantarum circa Cantabrigiam nascentium* (Cambridge, 1660), pp. 100–3.

65 Sprague, *The Evolution of Botanical Taxonomy*, pp. 16–17. See also Sachs, *History*, p. 61.

66 Sprague, *The Evolution of Botanical Taxonomy*, pp. 18–20.

67 *Ibid.*, p. 19.

68 Ray, *Catalogus plantarum circa Cantabrigiam nascentium*, pp. 100–3. Jung's notes reached Hartlib sometime in the 1650s. In 1660 Ray reports that he was given a copy of these notes by Dr Worthington who had received them from Hartlib

(Charles Raven, *John Ray, Naturalist, His Life and Works* (Cambridge, 1942), p. 106). John Beale communicated Jung's system to the Royal Society in May, 1663 ('Jung', *Dictionary of Scientific Biography*, VII, 194).

69 Ray, *Catalogus plantarum circa Cantabrigiam nascentium*, pp. 100–3.

70 *Ibid.*, p. 103.

3. Nomenclature

1 Michel Foucault, *The Order of Things: An Archaeology of the Human Sciences* (London, 1970), p. 159.

2 Brent Berlin, 'Speculations on the growth of ethnobotanical nomenclature', *Language in Society*, I (1972), p. 53.

3 *Ibid.*, p. 73.

4 *Ibid.*, p. 72.

5 *Ibid.*, pp. 73–8.

6 *Ibid.*, p. 78.

7 E. L. Greene, *Landmarks of Botanical History*, Smithsonian Miscellaneous Collections (Washington, D.C., 1909), 54, p. 110.

8 Stephen Ullmann, 'Semantic universals', in *Universals of Language*, ed. J. H. Greenberg (Cambridge, Ma., 1963), p. 181.

9 See Greene, *Landmarks*, pp. 107–11 for the way in which Theophrastus chose terms for trees, shrubs, undershrubs and herbs.

10 Berlin, 'Speculations', p. 72.

11 Claude Lévi-Strauss, *The Savage Mind* (1962; rpt. Chicago, 1968), pp. 153–4.

12 John Ray, 'A Discourse on the Specific Differences of Plants' (1674) in *Further Correspondence of John Ray*, ed. R. W. T. Gunther (London, 1928), pp. 77–83.

13 Berlin, 'Speculations', p. 74.

14 Goody, *The Domestication of the Savage Mind* (Cambridge, 1977); Havelock, *Preface to Plato* (Cambridge, Ma., 1963) and *Origins of Western Literacy* (Toronto, 1976); Ong, *Interfaces of the Word: Studies in the Evolution of Consciousness and Culture* (Ithaca, N.Y., 1977).

15 Bohuslav Havránek, 'The functional differentiation of the standard language' (Prague, 1932), in *A Prague School Reader on Esthetics, Literary Structure, and Style*, ed. and trans. Paul L. Garvin (Washington, D.C., 1964), p. 6.

16 *Ibid.*, pp. 6–7.

17 Berlin, 'Speculations', pp. 54–5.

18 *Ibid.*, p. 83.

19 Herbert E. Clark and Eve V. Clark, *Psychology and Language* (New York, 1977), p. 530.

20 Berlin, 'Speculations', p. 83.

21 See R. F. Jones, *The Triumph of the English Language* (1953; rpt. Palo Alto, Ca., 1966).

22 John Bullokar, *An English Expositor* (1616); Robert Cawdrey, *A Table Alphabetical* (1604); Henry Cockeram, *The English Dictionaries* (1623); Thomas Blount, *Glossographia* (1656); Edward Phillips, *The New World of English Words* (1658). There were also a number of specialized dictionaries; on this see Jones, *Triumph*, pp. 274–5. For full bibliographical details see Alston's volume on dictionaries (V)

in his *Bibliography of the English Language from the Invention of Printing to the year 1800* (Leeds, 1965–73).

23 Ralph Lever, *The Arte of Reason, rightly termed, Witcraft, teaching a perfect way to argue and dispute* (London, 1573); Arthur Golding, trans., *A Woorke concerning the trewnesse of the Christian Religion* by Philip de Monnay (London, 1587), p. 223; John Hart, *A Methode or comfortable beginning for all unlearned, whereby they may bee taught to read English, in a very short time, with pleasure* (London, 1570), Preface.

24 Gerard, *The Herbal or General Historie of Plants* (London, 1597), p. 1.

25 The situation was the same for chemistry in the late seventeenth and eighteenth centuries. For a detailed study of the inadequacy and the development of chemical nomenclature see Maurice P. Crossland, *Historical Studies in the Language of Chemistry* (Cambridge, Ma., 1962).

26 Ray, *Catalogus plantarum circa Cantabrigian nascentium* (Cambridge, 1660), pp. 34–7.

27 Blanche Henrey, *British Botanical and Horticultural Literature before 1880* (Oxford, 1976), I.

28 On the function of a name, see Ian Michael, *English Grammatical Categories and the Tradition to 1800* (Cambridge, 1970), passim and G. A. Padley, *Grammatical Theory in Western Europe 1500–1700* (Cambridge, 1976), passim.

29 Linnaeus, *Systema naturae*, 12th edn (Holmiae, 1776), p. 13.

30 Berlin et al., 'General principles of classification and nomenclature in folk biology', *American Anthropologist*, 75 (1973), pp. 216–17.

31 *Ibid.*, p. 217.

32 Frans Stafleau, *Linnaeus and the Linnaeans* (Utrecht, 1971), pp. 103–4.

33 Berlin, 'Speculations', pp. 72–3.

34 James Larson, *Reason and Experience* (Berkeley, Ca., 1971), p. 123.

35 W. T. Stearn, 'The background of Linnaeus' contribution to the nomenclature and methods of systematic biology', *Systematic Zoology*, 8 (1959), pp. 4–22.

36 Linnaeus, *Species plantarum* (Holmiae, 1753), I, 82. See John L. Heller, 'The early history of binomial nomenclature', *Huntia*, I (1964), p. 42.

37 A. Louis, 'Les Dénominations binominales des plantes dans l'antiquité grecque', *Scientiarum Historia*, 13 (1971), pp. 1–26, traces the evolution of binomial names in ancient Greek texts. He shows that, in Homer, of the 53 plant species named, only 3 have binomial names (5%). This is increased in the work of Hippocrates. Theophrastus lists 358 species out of which 66 have binomial names (18%). By the time of Dioscorides binomial naming is conscious and habitual (52%) with the species name forming a short description.

38 C. Váczy, 'Les Origines et les principes du développement de la nomenclature binaire en botanique', *Taxon*, 20 (1971), p. 580. See Theophrastus, *Historia plantarum*, 3.13.7.

39 C. Váczy, 'Les origines', p. 583. The examples are from *Historia plantarum* 7.4.4; 4.11.1; 7.4.5.

40 Caspar Bauhin, *Pinax theatri botanici*, 2nd edn (Basel, 1671), pp. 153–4.

41 *Ibid.*, p. 154.

42 *Ibid.*, p. 158.

43 Ray, *Catalogus plantarum circa Cantabrigiam nascentium*, pp. 34–5.

4. The groundwork

1 Leonard Bloomfield, *Language* (1933; rpt. New York, 1965), p. 21.

2 E. D. Hirsch, *The Philosophy of Composition* (Chicago, 1977), p. 18. The relation between written and spoken language is a complicated one. Readers who are interested in this subject are referred to the following: E. Haugen, 'Linguistics and language planning', in *Sociolinguistics*, ed. W. Bright (The Hague, 1966), pp. 50–67; F. Householder, *Linguistic Speculations* (Cambridge, Ma., 1971), ch. 13; Joseph Vacheck, 'Writing and phonetic transcription', in *Readings in Linguistics II*, ed. Eric Hamp et al. (Chicago, 1966), pp. 152–7.

3 Henry Bradley, 'On the relations between spoken and written language with special reference to English', A paper read at the International Historical Congress, April 1913 (London, n.d.), p. 5.

4 *Ibid.*, p. 9.

5 *Ibid.*, p. 2.

6 Vivian Salmon, rev. art. of Noam Chomsky's *Cartesian Linguistics* in *Journal of Linguistics*, 5 (1969), pp. 165–87; G. A. Padley, *Grammatical Theory in Western Europe 1500–1700* (Cambridge, 1977).

7 Aristotle, *On Interpretation*, trans. Harold P. Cook (Loeb Classical Library: Cambridge, Ma., 1961), Ia5–8.

8 Padley, *Grammatical Theory*, pp. 11–13.

9 *Ibid.*, pp. 154–84.

10 Francis Bacon, *De dignitate et augmentis scientiarum* in *The Works of Francis Bacon*, ed. and trans. James Spedding and Robert Ellis (1870; rpt. New York, 1968), IV, 439. Abbreviated hereafter as Sp.

11 Sp. IV, 439–40.

12 *Ibid.*

13 Francis Bacon, *Novum organum* (I. lix) in Sp. IV, 61. See also *De augmentis*, Sp. IV, 433–4.

14 *Ibid.*, I.lx, p. 61.

15 *Ibid.*

16 *Ibid.*, I.xv, p. 49.

17 *Ibid.*, I. xvi, p. 49.

18 *Ibid.*, I.lx, p. 62.

19 R. F. Jones, *Ancients and Moderns*, 2nd edn (1961; rpt. Berkeley, 1965), p. 49.

20 See Popper, *Conjectures and Refutations: The Growth of Scientific Knowledge* (1963; rpt. New York, 1968), pp. 12–18.

21 *Novum organum*, I.xiv, Sp. IV, 49. See also *De augmentis*, Sp. IV, 411.

22 *Ibid.*, I.lxiii, p. 64.

23 *Ibid.* Bacon himself was not above categorizing, as can be seen in the catalogue of particular histories ('Catalogus historiarum particularium') he appended to *Parasceve ad historiam naturalem et experimentalem*. Included here is a traditional Aristotelian taxonomy of natural substances. *The Advancement of Learning* is also a classification, albeit along different lines. Another point to note here is the connection between these classifications and encyclopedias, i.e. the classifications are seen as a way to exhaust all possible topics and hence to provide the framework for an

encyclopedic attempt to record all knowledge. This, to some extent, accounts for the Baconian influence on later figures in the encyclopedia tradition, e.g. Alsted and Comenius. On Bacon as an encyclopedist see Virgil K. Whitaker, 'Francis Bacon's intellectual milieu', in *Essential Articles for the Study of Francis Bacon*, ed. Brian Vickers (Hamden, Conn., 1968), pp. 28–50. See also 'Learned experience or the hunt of pan', in *De augmentis*, Sp. IV, 413 ff.: 'what really amazes the reader is the enormous proliferation of categories under which observed phenomena related to a given problem should be arranged or cross-classified in elaborate and endless tables' (Whitaker, p. 37).

24 *Novum organum*, I. lxxxvi, Sp. IV, 85.

25 *Ibid.*, II.v, p. 122.

26 *Ibid.*

27 *Ibid.*, II.xi, p. 127.

28 *Ibid.*, II.xii, pp. 129–32.

29 *Ibid.*, II.xiii, p. 137.

30 *Ibid.*, II.xv, p. 145.

31 *Ibid.*, II.i, p. 119.

32 See Paolo Rossi, *Francis Bacon*, trans. Sacha Rabinovitch (Bari, 1957; 1st English edn, London, 1968), pp. 194–201; A. C. Crombie, *Augustine to Galileo* (Harmondsworth, 1969), II, 289–96; Meyrick H. Carré, *Phases of Thought in England* (Oxford, 1949), pp. 244–5; Robert Kargon, *Atomism in England from Hariot to Newton* (Oxford, 1966), pp. 48–52; Lisa Jardine, *Francis Bacon: Discovery and the Art of Discourse* (Cambridge, 1967), pp. 109–14; Collingwood, *The Idea of Nature* (Oxford, 1965), pp. 100–1; Fulton H. Anderson, *The Philosophy of Francis Bacon* (Chicago, 1948), p. 160; Mary B. Hesse, 'Francis Bacon's Philosophy of Science', in *A Critical History of Western Philosophy*, ed. D. J. O'Connor (New York, 1964), rpt. in *Essential Articles*, ed. Brian Vickers, pp. 114–39.

33 Jardine, *Francis Bacon*, p. 114.

34 *Novum organum*, II. xvii, Sp. IV, 146.

35 *Ibid.*, II. xx, p. 150. See Anderson, *Philosophy*, p. 160.

36 Crombie, *Augustine to Galileo*, II, 290.

37 Carré, *Phases*, p. 245.

38 Collingwood, *Idea of Nature*, p. 101.

39 *Novum organum*, II. xx, Sp. IV, 155.

40 For the life of Comenius, see G. H. Turnbull, *Hartlib, Dury and Comenius*. For Alsted, see Percival Cole; *A Neglected Educator: Johann Heinrich Alsted* (Sydney, 1910) and Hans Aarsleff, 'Alsted' in *The Dictionary of Scientific Biography*, ed. C. C. Gillispie (New York, 1970–80), I, 359–63. For Bisterfeld, see Paolo Rossi, *Clavis universalis* (Milan, 1960), pp. 197–200 and *The Dictionary of Scientific Biography*, II, 163–4.

41 In Cole, *Neglected Educator*, p. 23. See also Leroy E. Loemaker, 'Leibniz and the Herborn Encyclopedists', *Journal of the History of Ideas*, 22 (1961), pp. 323–38. Alsted's encyclopedia was published under the title *Scientiarum omnium encyclopaediae* (Herborn, 1630).

42 John Amos Comenius, *The Great Didactic*, ed. and trans. M. W. Keatinge (London, 1896), pp. 230–321. This was originally written in Czech between 1627 and 1632

but was first published, in Latin, in *Opera omnia* (Amsterdam, 1657) under the title *Didactica magna.*

43 *Ibid.*, p. 267.

44 Comenius, *Continuatio admonitionis fraternae* (Amsterdam, 1669) in Robert Young, *Comenius in England* (London, 1932), pp. 30–1.

45 In 1638 Hartlib published Comenius' *Conatuum Comenianorum praeludia* at Oxford. Comenius then published *Conatuum pansophicorum dilucidatio* (Leszno, 1638). In 1639 Hartlib published a reprint of this under the title *Pansophiae prodromus.* In 1642 he then published an English translation entitled *A Reformation of the Schools.* These quotations are from 'A Dilucidation...' in *A Reformation of the Schools* (London, 1642), p. 64.

46 *Linguarum methodus novissima* (Leszno, 1648) in Comenius, *Analytic Didactic*, ed. Vladimir Jelinek (Chicago, 1953), p. 38.

47 Comenius, *Via lucis*, ed. and trans. E. T. Campagnac (Liverpool, 1938). The work was written in 1641 and circulated in manuscript in England; it was published in Amsterdam in 1668.

48 See Comenius, *A Pattern of Universal Knowledge...or a Diatyposis or Model of the Eminently Learned and Pious Promoter of Science in Generall*, trans. Jeremy Collier (London, 1662), p. 15. Originally published as *Pansophiae diatyposis* (Danzig, 1643).

49 *Ibid.*, p. 22.

50 *Janua linguarum reserata* (Leszno, 1631). A pirated edition (in French/English/Latin) was published in London in 1631 by John Anchoran under the title *Porta linguarum trilinguis reserata et aperta, sive seminarium linguarum & scientiarum omnium: The Gate of Tongues unlocked and opened or else a seminarie or seed-plot of all tongues and sciences.* This has been reproduced in facsimile by Scolar Press. The English/Latin *Janua* later became known as the *Janua linguarum reserata sive omnium scientiarum & linguarum seminarium: The Gate of Tongues unlocked or a Seedplot of all Arts and Tongues.*

51 Comenius, *The Great Didactic*, p. 22. Young, *Comenius*, p. 4, points out that here Comenius writes in the spirit of Raymond Lull's thirteenth-century works, *Ars generalis* and *Ars brevis*, as well as Alsted's *Clavis ars Lullianae* (1633). Salmon notes that the classification of words into groups had been customary with English grammarians of Latin, e.g. Hunter and Stanbridge (Vivian Salmon, 'Language planning in 17th century England: its contexts and aims', in *In Memory of J. R. Firth*, ed. Charles F. Bazell (London, 1966), p. 381).

52 Comenius, *Janua linguarum reserata vestibulum* (Leszno, 1633). This was added to the 4th edition of the English *Janua* in 1633. I have used the 6th edition, *Janua linguarum reserata: The Gate of Languages Unlocked.* Formerly translated by T. Horn; corrected and amended by J. Robotham. Together with a *Portal to the Janua* (London, 1643).

53 Comenius, *Orbis sensualium pictus* (Nuremburg, 1658). *The Visible World: or A picture and nomenclature of all the chief things that are in the world; and of men's employments therein*, trans. Charles Hoole, 3rd edn (London, 1672), intro. James Bowen (Sydney, 1967), author's preface, A3v.

54 *Ibid.*

55 *Ibid.*, A6r.

56 Comenius, *Natural Philosophy Reformed by Divine Light or a Synopsis of Physics* (London, 1651), pp. 79–81. Originally published as *Physicae ad lumen divinum reformatae synopsis* (Leipzig, 1633).
57 Comenius, *Janua rerum reserata sive universalis sapientia*, ed. Klaus Schaller (Munich, 1968). It was written between 1640 and 1641 and published posthumously in Leyden in 1681.
58 Comenius, *Via lucis*, p. 137.
59 *Ibid.*, p. 144.
60 *Ibid.*, pp. 146–8.
61 *Ibid.*, pp. 152–3.

5. Proposals and schemes for a universal language

1 Hugh Trevor-Roper, 'Three foreigners: the philosophy of the puritan revolution' in Trevor-Roper, *Religion, Reformation and Social Change*, 2nd edn (London, 1972), pp. 258, 289. See also R. F. Jones, *Ancients and Moderns* (Berkeley, Ca. 1965).
2 Charles Webster, 'Marcaria: Samuel Hartlib and the Great Reformation', *Acta Comeniana*, 2 (1971), p. 159.
3 The fullest account of Hartlib's life is to be found in G. H. Turnbull, *Hartlib, Dury and Comenius* (Liverpool, 1947). See also Trevor-Roper, 'Three foreigners'.
4 See Hartlib, *Samuel Hartlib and the Advancement of Learning*, ed. Charles Webster (Cambridge, 1970); Vivian Salmon, 'Problems of language teaching; a discussion among Hartlib's friends', *Modern Language Review*, LIX (1964), pp. 13–24; and Charles Webster, *The Great Instauration: Science, Medicine and Reform 1626–1660* (New York, 1976), passim. On the clausary method of language teaching see Salmon, 'Joseph Webbe: some seventeenth century views on language teaching and the nature of meaning', *Bibliothèque d'Humanisme et Renaissance*, 23 (1961), pp. 324–40.
5 A. Rupert Hall, 'Science, technology and utopia', in *Science and Society 1600–1900*, ed. Peter Mathias (Cambridge, 1972), p. 41.
6 Charles Webster, 'English medical reformers of the Puritan Revolution', *Ambix*, XIV (1967), pp. 16–41; C. Webster, *The Great Instauration*, pp. 246–323; *Medicine in England*, ed. Allen Debus (Berkeley, Ca., 1974), passim; P. M. Rattansi, 'Paracelsus and the Puritan Revolution', *Ambix*, XI (1963), pp. 24–32; Debus, *The English Paracelsians* (New York, 1965), esp. chapters 3 and 4. On the Rosicrucian connection see Yates, *Rosicrucian Enlightenment* (London, 1972), pp. 197–234.
7 See Edmond FitzMaurice, *The Life of Sir William Petty* (London, 1895).
8 The name refers to Hartlib's plan for an Office of Address, a sort of college, museum and society which would link contemporary English life with the scientific and technological movements of the day. Yates argues that this group had Rosicrucian and hermetic origins (*Rosicrucian Enlightenment*, pp. 211–34). See also Hall, 'Science, technology and utopia', in *Science and Society*, ed. Mathias, pp. 33–53; R. H. Syfert, 'The origins of the Royal Society', *Notes and Records of the Royal Society*, 5 (1948), pp. 75–137; Margery Purver, *The Royal Society: Concept and Creation* (London 1967), pp. 194–205.

9 J. R. Jacob, *Robert Boyle and the English Revolution* (New York, 1977); H. F. Kearney, 'Puritanism, capitalism and scientific revolution', in *The Intellectual Revolution of the Seventeenth Century*, ed. Charles Webster (London, 1974), pp. 237ff.; P. M. Rattansi, 'The social interpretation of science', in *Science and Society*, ed. Mathias, pp. 18ff.

10 Barbara Shapiro, *John Wilkins: An Intellectual Biography 1614–1672* (Berkeley, Ca., 1969), pp. 14ff.

11 *Ibid.*, pp. 19–20.

12 On Haak's life see Pamela Barnett, *Theodore Haak, F.R.S.* (The Hague, 1962).

13 Hans Aarsleff, 'Wilkins', *Dictionary of Scientific Biography*, XIV, 366–8.

14 Yates, *Rosicrucian Enlightenment*, p. 226.

15 *Ibid.*, p. 226.

16 13 July 1654, John Evelyn, *The Diary of John Evelyn esq., F.R.S.*, ed. William Bray (London, 1906), I. 295,

17 Aarsleff, 'Wilkins', p. 366–8.

18 According to John Wallis, Haak was the founder of the informal scientific society which began to meet in London in 1645, which included among its members Wilkins, Ward and Wallis. According to Wallis it was this group which formed the nucleus of what later became the Oxford scientific 'clubb' although this view is rejected by some scholars. See R. H. Syfert, 'The origins of the Royal Society'; Pamela Barnett, 'Theodore Haak and the early years of the Royal Society', *Annals of Science*, XIII (1957), pp. 205–18; G. H. Turnbull, 'Samuel Hartlib's influence on the early history of the Royal Society', *Notes and Records of the Royal Society*, X (1953), pp. 101–30; Purver, *The Royal Society*, passim.

19 On 'natural magic' see Keith Thomas, *Religion and the Decline of Magic* (Harmondsworth, 1973), pp. 264–74.

20 James Knowlson, *Universal Language Schemes in England and France 1600–1800* (Toronto, 1975), p. 46.

21 There is some question about the identity of the author of this placard. For details see Knowlson, *Universal Language Schemes*, pp. 48–9. Des Vallées is mentioned by Charles Sorel, *De la Perfection de l'homme* (Paris, 1655), p. 346, cited by Cornélis de Waard, ed. *Correspondence du P. Marin Mersenne* (Paris, 1932), II, 329.

22 Descartes to Mersenne, 20 November 1629 in *Oeuvres Complètes de Descartes*, ed. Charles Adams and Paul Tannery (Paris, 1897), II, 77–82. For an English translation see *Descartes' Philosophical Letters*, ed. A. Kenny (Oxford, 1970), pp. 3-6. Tannery's notes record that according to A. Baillet, *La Vie de Monsieur Descartes* (Paris, 1691), II, 485, Christopher Wren produced an essay on a universal language based on Descartes' ideas. No work of this sort has been traced, although Wren did participate in discussions of Wilkins' language projects.

23 Mersenne to Haak, September 1640 and 16 November 1640 in Mersenne, *Correspondance*, XI, 417–18 and 422–3. See also James Knowlson, 'Jean Le Maire, the Almérie, and the "Musique Almérique"', *Acta Musicologica*, 40 (1968), pp. 86–9. Haak is said to have obtained the patent for Le Maire's device. A spin-off from Le Maire's work is that of Charles Sorel, reportedly working on a universal grammar around 1635 (of which there is no known publication). Sorel includes comments on this subject in his encyclopedia, *La Science universelle* (Paris, 1644),

IV, cited by Salmon, rev. art. of Noam Chomsky's *Cartesian Linguistics* in *Journal of Linguistics*, 5 (1969), p. 171. In his discussion, Sorel refers to the works of Alsted. See also p. 234, n. 21.

24 Bodleian Library, Ashmolean MS 788, f. 51, c. 1627–8.

25 *Ibid.*

26 Hartlib, *Ephemerides*, Sheffield University Library, 1635 BC 4.

27 *Ibid.*, 1639 BC 8. 'Sir Wm. Boswell affirms that Champagnolles having had much to deal with the man were able to perform what he promised and that he had done sọe l. book of Homers Odyssey which he had printed.' Sir William Boswell, English Ambassador to the Hague, member of Gresham College, and long time friend of Hartlib was often involved in gathering information on the universal language projects for Hartlib.

28 *Ibid.*, 1650 FG 8. See also for that year BC 3: 'Champagnolles' widow lives in Fenchurch Street and is one of the patients of Dr Gurdain. Sir W. Boswell affirms to have his art but bound up to secrecy as long as he lived, the whole direction being contained in half a sheet on both sides which he may now lawfully publish.' BC4: 'Dr Gurdain hath spoken with Champagnolles' widow and son who say that the new invented art of his characters are safe in their keeping, having been sealed up these 7 years. And being told that Sir Wm. Boswell knew the art, they replied that indeed he had one sheet of it, but that there were many other things or writings belonging to it which he had not. They also told the doctor that 2 persons great lovers of art and ingenuity had been written in France and Low Countries and till they had returned their answers of accepting or refusing, they could not transact with anybody else. But promised to conclude nothing however without first acquainting the doctor.' The doctor mentioned, Dr Aaron Gourdain, was, like Petty, Ward, Boswell and a host of others, involved in Paracelsian medicine.

29 *Ibid.*, 1650 FG 8. It would seem likely that it was Champagnolles' scheme that Seth Ward had in mind when he referred to a character 'in manuscript shewed to King Charles (containing the first book of Homer's Illiads)' (*Vindiciae academiarum*, Oxford, 1654, p. 21). And it may have inspired the verses preceding Cave Beck's *Universal Character* (London, 1657): 'Speech is the Index of the mind; Loe here/ Th'Index of Speech, the Dumb Interpreter;/The Illiad's in a Nut-Shell; Tongues in Brief;/Babel Revers'd; the traveller's Relief...' In 1694 Richard West writes 'Now we may Homer's Iliads confine/Not in a nut shell but a point or line.'

30 E. S. Shuckburgh, ed., *Two Biographies of William Bedell* (Cambridge, 1902), p. 110. Bedell was strongly sympathetic with the movement towards unification of Protestant churches and to this end contributed support to Hartlib's colleague, John Dury.

31 In 1641 Hartlib records that 'Mr. Flower, Fellow of Jesus College in Oxford knows fully how to write to Mr. Johnson in Ireland, author of Real Characters' (*Ephemerides*, 1641 A4). In 1639 Hartlib records that Johnson 'purposes to grave his characters next year 1640' (JJ 8). In 1640 'Sir W. [Boswell] has heard that he [Johnson] hath first printed some sheets of his Real Character but they will hardly be done with in these 2 years. It may be that he is to deale for the cutting of them with some in the Low Country' (BC 8). See Mersenne, *Correspondance*, XI, 418.

32 Hartlib, *Ephemerides.*

In 1647 Haak wrote to Mersenne of what is probably Johnson's project: 'we are still awaiting a gentleman from Ireland from whom we are told to expect great things, as he brought to conclusion so I am informed, the efforts of more than twenty years of a certain very skillful man of that kingdom (died 3 or four years ago) to write in a way that the message may be read at once in all languages known or understood' (Haak to Mersenne, 24 May/3 June 1647, Bibliothèque Nationale MS NAF 6206, f. 89 in Knowlson, *Universal Language*, pp. 53–4).

In 1649 Hartlib entered into his *Ephemerides* (AB 8) that 'Such an occasion is in Sumner's offer for bringing over Johnson's real char[acter]. See his letter and Bond's paper.' In 1650 business was still going forth on Johnson's character. Hartlib reports that 'Mr. French, an Englishman somewhat of a cynical disposition, but very mathematical and of strong imaginations, will be a very fit man amongst others to perfect Johnson's Real Characters' (NJ 8). Mr French is not identified. He is most likely John French, protégé of Lord Saye and Paracelsian physician well known to Hartlib's circle. Hartlib also mentions here William Dugard, headmaster of the Merchant Taylors' School who helped publish the works of Hartlib and Comenius and was a translator of Paracelsus. In 1650, Hartlib goes on to record that Fuller 'found fault with Johnson's characters for making characters for all synonyms, and was of the opinion that his later contrivances and labours were far beyond his first thoughts; of which this draught seemed to be which his wife the widow hath brought over' (*Ephemerides*, KL 1). This is probably Thomas Fuller who lived in Ireland and was consecrated Bishop of Ardfert in 1641. He fled to London at the time of the Rebellion and lived in Oxford between 1645 and 1660 (*Dictionary of National Biography* ed. Sir Leslie Stephen and Sir Sidney Lee (Oxford, 1921–2), VII, 760). The information about Fuller was reported to Hartlib by Dr Richardson. This is probably John Richardson who, in 1633, succeeded Bedell as Bishop of Ardagh. He too fled to England in 1641 and died there in 1654 (*Dictionary of National Biography* XVI, 1114). Hartlib also mentions that 'Dr Fuller with 2 more have designes to perfect an art of real characters from Dr. Richardson's relation.'

33 Francis Godwin, *The Man in the Moone: or a Discourse of a Voyage Thither by Domingo Gonsales, the Speedy Messenger* (1638), ed. Grant McColley in *Smith College Studies in Modern Languages*, 19 (1937), pp. 1–48. According to McColley, the tale was probably written c. 1630. See also H. Neville Davies, 'Godwin's "Lunatique Language"', *Journal of the Warburg and Cortauld Institute*, 30 (1967), pp. 296–316 and James Knowlson, 'A note on Bishop Godwin's *Man in the Moone:* the East Indies Trade Route and a 'language of musical notes"', *Modern Philology*, 65 (1965), pp. 357–61.

34 Godwin, *Man in the Moone*, pp. 35–6.

35 *Ibid.*, p. 36.

36 *Nuncius inanimatus* (London, 1629). This was translated as *Nuncius inanimatus or the Mysterious Messenger* by Thomas Smith. The second edition of 1657 is reprinted by McColley in *Smith College Studies in Modern Languages*, 19 (1937), pp. 50–67.

37 See Knowlson, 'A note on... *Man in the Moone*', p. 360; Paul Cornelius, *Languages in Seventeenth- and Eighteenth-Century Imaginary Voyages* (Geneva, 1965),

45–8, 51; E. Seeber, 'Ideal languages in the French and imaginary voyages', *Publications of the Modern Language Association*, LX (1945), pp. 568ff. Imaginary voyages were the source of several imaginary languages, e.g. More's *Utopia* (1516) and Rabelais' *Pantagruel;* Cyrano de Bergerac's *Voyage dans la lune* (c. 1650) also has a musical language for either voices or instruments.

38 John Wilkins, *Mercury; Or, the Secret and Swift Messenger: Shewing How a Man May with Privacy and Speed Communicate His Thoughts to a Friend at Any Distance* (London, 1641). Reprinted in *The Mathematical and Philosophical Works of the Right Reverend John Wilkins* (London, 1802), II.

39 Wilkins had read Godwin's *Man in the Moone* shortly after its publication in 1638. This had led him to read *Nuncius inanimatus* which, as he indicates in the Preface to *Mercury* (p. 17), was the source for *Mercury*. McColley speculates that Wilkins came to know of Godwin's works through John Norton, the printer of Godwin's *Man in the Moone* and the third edition of Wilkins' *Discovery* (1640). See McColley's introduction to *The Man in the Moone*, pp. x–xi.

40 *Mercury*, p. 55. All references are to the original edition.

41 *Ibid.*

42 *Ibid.*, pp. 54–6.

43 Comenius, *Via lucis* (Amsterdam, 1668), ed. and trans. E. T. Campagnac (Liverpool, 1938).

44 See Benjamin De Mott, 'Comenius and the real character in England', *Publications of the Modern Language Association*, LXX (1955), pp. 1071ff.; Salmon, 'Language planning', pp. 372ff.; Aarsleff, 'Wilkins', pp. 366–8.

45 See Chapter 6; Mersenne, *Questions thèologiques* (1634), p. 138; Robert Lenoble, *Mersenne ou la naissance du mécanisme* (Paris, 1943), p. 518.

46 Théophraste Renaudaut, *A General Collection of Discourses of the Virtuosi of France*, trans. G. Havers (London, 1664), pp. 253–4.

47 22 November 1640, Mersenne, *Correspondance*, X, 267.

48 For Bisterfeld, see *Dictionary of Scientific Biography*, II, 163–4; see also Mersenne, *Correspondance*, VII, 187–8 and 217–18 as well as VIII, 534.

49 Salmon, 'Language planning in 17th century England: its contexts and aims', in *In Memory of J. R. Firth*, ed. Charles F. Bazell (London, 1966), p. 293.

50 Mersenne, *Correspondance*, VIII, 575–7 and 574–80; Aarsleff, 'Wilkins', p. 368.

51 Mersenne, *Correspondance*, XI, 412; IX, 16, 42 and XI, 417–18, 422–3; XI, 418.

52 Salmon, 'Language planning', pp. 391–2; Mersenne, *Correspondance*, X, 263ff.

53 Comenius, *Via lucis*, p. 8.

54 *Ibid.*, pp. 183–4.

55 *Ibid.*, p. 191.

56 *Ibid.*, p. 219.

57 *Ibid.*

58 Hartlib, *Ephemerides*, 1640 AB 4 quoting Joachim Hubner, a refugee from Cleves, assistant to Comenius and correspondent of Hartlib who was known to have been interested in the Rosicrucians.

59 *Ibid.*, 1639 CD 7.

60 *Ibid.*, 1639 DE 3.

61 *Ibid.*, 1639 EF 8.

62 Francis Lodowyck, *A Common Writing: Whereby two although not understanding one another's Language, yet by the helpe thereof may communicate their minds to one another* (London, 1647) and *The Groundwork and Foundation Laid (or so Intended) For the Framing of a New Perfect Language: And an Universal or Common Writing* (London, 1652). They are both reprinted in *The Works of Francis Lodwick: A study of his writings in the intellectual context of the 17th Century*, ed. Vivian Salmon (London, 1972). Salmon provides an extensive and excellent introduction. All references are to the original editions.

63 Lodowyck, 'Concerning Mr. Dalgarno's universall caracter and language', (1657) British Museum Sloane MS 932, ff. 13r–15v and 'Of an universall reall caracter', (1657) Sloane MS 897, ff. 32r–39v. Both are reprinted in *The Works of Francis Lodwick*, ed. Salmon.

64 Lodowyck, 'An Essay Towards an Universal Alphabet and Universal Primer', *Philosophical Transactions of the Royal Society*, 16, no. 182 (1686), 126–33.

65 Lodowyck, *Common Writing*, p. 2.

66 *Ibid.*, pp. 2–3.

67 *Ibid.*, pp. 8–9.

68 Lodowyck, *The Groundwork*, p. 8.

69 *Ibid.*, pp. 8–9.

70 *Ibid.*, p. 8.

71 *Ibid.*, p. 9.

72 British Museum Sloane MS 897, f. 33 r–v, reprinted in *The Works of Francis Lodwick*, ed. Salmon. This list of possible grammatical terminations is more a set of notes than a finished treatment and is too long to be reproduced here.

73 Lodowyck, *Common Writing*, p. 28.

74 Lodowyck, *Groundwork*, p. 13.

75 *Ibid.*, p. 10.

76 Boyle to Hartlib, 19 March 1647, in *The Works of the Honourable Robert Boyle*, ed. Thomas Birch (London, 1772), I, xxxvii-xxxviii.

77 Boyle to Hartlib, 8 April 1647, in *The Works of...Boyle*, I, xxxix.

78 *Arithmeticus nomenclator, mundi omnes nationes ad linguarum et sermonis imitatem invitans; auctore linguae (quod mirere) Hispano quodam, vere, ut dicitur, muto.* The author has been identified as Pedro Bermudo, S. J. See Joaquin Carreras y Artau, *De Ramón Lull a Los Modernos Ensayos De Formación De Una Lengua Universal* (Barcelona, 1946), pp. 22–3. It was later described by Kircher's disciple, Gaspar Schott in *Technica curiosa* (Nuremberg, 1664), pp. 483ff. This may be the Spanish priest that Kenelm Digby refers to as curing a Spanish nobleman who was deaf and dumb (see Kenelm Digby, *Two treatises; In the one of which, the nature of bodies; in the other, the nature of man's soule, is looked into* (Paris, 1643), p. 254). M. H. Nicolson (*Conway Letters 1642–1684* (Oxford, 1930), p. 310) describes Francis Mercury Van Helmont, hermeticist and cabalist, as an instructor of the deaf and dumb, for whom he invented his Alphabet of Nature, and the contriver of a universal language. In 1654 he published *Alphabeti vere naturalis hebraici... delineatio quae simul methodum suppeditat, juxta quam qui surdi nati sunt sic informari possunt, ut non alios saltem loquentes intelligent, sed et ipsi ad sermonis usum perveniant* (Sulzbaci, 1657).

79 The list is clearly set out in Schott, *Technica curiosa*, pp. 485–96.
80 *Ibid.*, p. 404.
81 *Ibid.*, p. 501.
82 Hartlib, *Ephemerides*, 1653. This is no doubt Francis Potter, rector of Kilmington, Somerset. In 1642 he published 'On the Number 666' which Hartlib distributed in manuscript. Potter was also an inventor, Fellow of the Royal Society and friend of Aubrey. Aubrey records that Potter knew Wilkins 'very well' (John Aubrey, *Brief Lives*, ed. Oliver Dick (London, 1958), p. 478; see also C. Webster, *The Great Instauration: Science, Medicine and Reform 1626–1660* (New York, 1976), p. 33).
83 This is Bengt (or Benedict) Skytte, a Swedish nobleman and educator. Skytte's connections with Comenius and the Protestant reformers go back quite a way. His father, Johann Skytte, had known Bacon and, as Chancellor of Uppsala University, had been instrumental in bringing Comenius to work in Sweden. The family was active in circles for Protestant reconciliation and knew Hartlib's partner, John Dury when Dury was residing in Stockholm between 1636 and 1638. Bengt Skytte followed his father as an educator and statesman. Between 1659 and 1662 he lived in London as a private citizen. In 1659 he attended the London scientific meetings (with Wilkins). He formulated at this time plans for a scientific society somewhat akin to Hartlib's scheme for a Bureau d'Adresse. Because of their religious overtones neither of these schemes were utilized by the Wilkins group in forming the Royal Society. Skytte returned to Sweden in 1662 but in 1664 commenced a long period of travel, living the life of an adventurer and vainly seeking appointments. He returned to Stockholm not long before his death in 1683. (See Purver, *Royal Society*, pp. 220–1.)
84 Elias Ashmole, trans., *Fasciculus chemicus*, by Arthur Dee (London, 1650), Prolegomena.
85 Cave Beck, *The Universal Character, By which all Nations in the World may understand one another's Conceptions, Reading out of one Common Writing their own Mother Tongues* (London, 1657). The work was originally published under the title, *Le Charactere universel* (London, 1657) although there are no known surviving copies of this. Beck is also known to have written a treatise on memory; see Royal Society Classified Papers, 16, no. 2.
86 Nathaniel Fairfax to Henry Oldenburg, 17 July 1668 in Oldenburg, *Correspondence of Henry Oldenburg*, ed. A. Rupert Hall and Marie Boas Hall (Madison, Wis., 1968–), IV, 555.
87 Beck to Oldenburg, 15 August 1668, Royal Society Letter Book, BI, f. 137.
88 Beck, *Universal Character*, p. 11, pp. 29–30.
89 *Ibid.*, p. 18.
90 Dalgarno to Hartlib, 20 April 1657, British Museum Sloane MS 4377, f. 148r.
91 Anthony Wood, *Athenae Oxonienses*, 2nd edn (London, 1721), II, 506–7.
92 Morstyn to Hartlib, April 1657, Sloane MS 4377, f. 149. Salmon in *The Works of Francis Lodwick*, p. 63, reports that in 1653 a Dutchman named David Goubard or Goubart was working on a shorthand composed of primitive words or notions, with a character for a whole word, which could be read in any language.
93 Hartlib, *Ephemerides*, 1657.

94 Dalgarno to Hartlib, 20 April 1657, Sloane MS 4377, f. 148.
95 Dalgarno, 'A New Discovery of the Universal Character, Containing Also a More Readie and Approved Way of Short-hand Writting, then any heretofore practised in this Nation' (Oxford, 1657), Sloane MS 4377, f. 144. Reproduced in Alston, *The Bibliography of the English Language from the Invention of Printing to the Year 1800* (Leeds, 1965–73), VII.
96 Dalgarno, 'The Tables of the Universal Character' (Oxford, 1657), Sloane MS 4377, ff. 145–6. This is reproduced in the facsimile of *Ars signorum* published by Scolar Press.
97 Dalgarno, 'A New Discovery', Sloane MS 4377, f. 144.
98 Dalgarno to Hartlib, 20 April 1657, Sloane MS 4377, f. 148.
99 Hartlib to Pell, July 1657, Sloane MS 4377, f. 149.
100 Dalgarno to Pell, 12 December 1657, Sloane MS 4377, f. 150.
101 *Ibid.*, f. 151.
102 Dalgarno, 'Tables', Sloane MS 4377, f. 145.
103 Dalgarno to Pell, 12 December 1657, Sloane MS 4377, f. 151.
104 Vivian Salmon, 'The evolution of Dalgarno's "Ars signorum"', in *Studies in Language and Literature in Honour of Margaret Schlauch*, ed. M. Brahmer et al. (Warsaw, 1966), p. 359. Robert Wood was a Fellow of Lincoln College, Oxford and friend of Petty engaged in research for a universal language. See Hartlib's *Ephemerides*, July 1658: 'He was appointed by the Clerk to make a catalogue of matter for all Trades, Artificers, etc. in the Oxford Library, of which he hath collected some sheets.'
105 Salmon, 'Evolution', p. 359.
106 *Ibid.*
107 Hartlib, *Ephemerides*, 1657, 51–512.
108 Lodowyck, 'Concerning Mr. Dalgarno's Universall Carracter and Language', Sloane MS 932, ff. 13r–15v.
109 John Wallis, *A Defence of the Royal Society and Philosophical Transactions in Answer to the Cavils of Dr. William Holder* (London, 1678), p. 16.
110 Louis Couturat, *La Logique de Leibniz* (1901; rpt. Hildesheim, 1961), p. 59; James Knowlson, 'The ideal languages of Veiras, Foigny, and Tyssot de Patot', *Journal of the History of Ideas*, 24 (1963), p. 278.
111 Johann Joachim Becher, *Spirensis character pro notitia linguarum universali* (Frankfurt, 1661). See Somerset, Edward (Earl of Worcester), *A Century of Names and Scantlings of Such Inventions, As at present I can call to Mind to have tried and perfected* (London, 1665), ed. Charles F. Partington (London, 1825), Editor's note, pp. 34–6. On Becher's life see Worthington, *The Diary and Correspondence of Dr. John Worthington*, ed. J. Crossley (London, 1847), I, 228–31.
112 Becher, *Character*, B4, B5, B8r. See also Schott, *Technica curiosa*, pp. 518ff. In Johann Christopher Sturm, *Collegium experimentale sive curiosum* (Nuremberg, 1676) there is an essay on Becher's scheme and Schott's modifications. Sturm's book was announced in the *Philosophical Transactions of the Royal Society* on 24 January 1676 (10, 509–12).
113 Schott, *Technica curiosa*, pp. 518ff.
114 The form of these is taken from coin boards or number tables and reckoning

boards popularized in Europe in the 16th century; see Karl Menninger, *Number Words and Number Symbols*, trans. Paul Broneer (Cambridge, Ma., 1969), pp. 340–53. In 1681 Joachim Frischius of Riga was reported to be engaged on a language scheme similar to Becher's. A specimen was printed at Thorn in 1681 but apparently has not survived. See Somerset, *A Century of Names*, p. 36,

115 Athanasius Kircher, *Polygraphia nova et universalis ex combinatoria arte detecta* (Rome, 1663).

116 His Cabalist writings (1663) were known to Ward. In 1669 Kircher published *Ars magna sciendi* reputed to be a modification of Lull's combinatorial method. The base of Lull's scheme is a table of six categories (*subjecta*, etc.) each of which is broken into nine classes (i.e. Deus, Angelus, Coelum, Homo, Imaginativa, Sensitiva Vegetativa, Elementativa, Instrumentativa). A letter of the alphabet represented the classes, one through nine. Kircher replaced these letters so that each class in each category had a unique sign, e.g. △ Deus, ◎ Coelum, □ Elementa, ⁞⁞⁞ Instrumentiva (from Louis Couturat and L. Leau, *Histoire de la langue universelle* (Paris, 1903), pp. 541–2).

117 On Kircher's life and works see the *Dictionary of Scientific Biography*, VII, 374–8.

118 Walter Pope to Abraham Hill, 12 November 1663, in Abraham Hill, *Familiar Letters* (London, 1767), p. 85. Walter Pope was Wilkins' half-brother; Sir Robert Moray, the first President of the Royal Society.

119 Kircher, *Polygraphia*, p. 12.

6. Proposals for a philosophical language

1 See above, pp. 93–7.

2 *Descartes' Philosophical Letters*, ed. Anthony Kenny (Oxford, 1970), pp. 3–6. The original is in Descartes, *Oeuvres*, II, 76–82. Leibniz' comments on Descartes' letter are given in Louis Couturat, *Opuscules et fragments inédits de Leibniz* (Paris, 1903), p. 28.

3 *Ibid.*, pp. 5–6.

4 *Ibid.*, p. 6.

5 Descartes, *Philosophical Works of Descartes*, trans. E. Haldane and G. Ross (Cambridge, 1931), I, 14.

6 Harold Joachim, *Descartes's Rules for the Direction of the Mind*, ed. Errol Harris (London, 1957), pp. 26–7.

7 *Ibid.*, p. 34.

8 *Ibid.*, p. 24.

9 *Ibid.*, p. 71.

10 See Brian E. O'Neil, *Epistemological Direct Realism in Descartes' Philosophy* (Albuquerque, N.M., 1974), p. 76.

11 Marin Mersenne, *Harmonie universelle* (Paris, 1936), Livre Premier de la Voix, Propositions XLVII–L, pp. 65–79. See Salmon, 'Language planning in 17th century England: its contexts and aims', in *In Memory of J. R. Firth*, ed. Charles F. Bazell (London, 1966), pp. 373 ff. and Knowlson, 'The ideal languages of Veiras, Voigny, and Tyssot de Patot', *Journal of the History of Ideas*, 24 (1963), pp. 275ff.

12 Mersenne, *Harmonie universelle* (Paris, 1636), Livre premier de la voix, Proposition XLVII, pp. 65–70.

13 *Ibid.*, Proposition XLIX, p. 73.

14 Mersenne also suggests using the octave as a real character: 'celuy qui cognoistroit toutes les especes des plantes, des animaux, des mineraux & des pierres, pourroit les exprimer & enseigner toutes les sciences avec toutes sortes d'instrumens de Musique' (*Harmonie Universelle*, Livre premier de la nature et des proprietez du son, Proposition XXI, p. 41).

15 *Harmonie Universelle*, Livre premier de la voix, Proposition XLIX, pp. 73–4.

16 *Ibid.*, Proposition XLVII, p. 65.

17 *Ibid.*, Proposition L, pp. 75–6.

18 Mersenne to Peiresc, 20 April 1635, *Correspondance*, V, 136.

19 See above, pp. 112–14.

20 G. H. Turnbull, *Hartlib, Dury and Comenius* (Liverpool, 1947), pp. 382ff.

21 Cyprian Kinner, *A Continuation of J. A. Comenius School Endeavours* (London, 1648), To the Reader.

22 *Ibid.*, pp. 1–2.

23 *Ibid.*, p. 2.

24 *Ibid.*, p. 3.

25 *Ibid.*

26 *Ibid.*, p. 4.

27 *Ibid.*, p. 5.

28 Benjamin De Mott, 'The sources and development of John Wilkins' philosophical language', *Journal of English and Germanic Philosophy*, LVII (1958), pp. 1–3.

29 We should remember here that Mersenne also was interested in an artificial language for botany as he told Haak in 1640. Since we do not know when Kinner was teaching it is difficult to tell what his relations with Mersenne were. If Kinner is referring to the period in which he worked with Bisterfeld and Alsted (1634–5) then Bisterfeld could have relayed this information to Mersenne. On this score it might be worth noting that when Kinner was drawing up his plans for his *Continuation* he expected help from Joachim Jung, the logician and botanist who had made contributions to plant taxonomy (see above, pp. 60–2). On the other hand it is not impossible that Kinner learned of Mersenne's notions for a botanical nomenclature from Comenius or his various associates.

30 De Mott, 'Sources and development', p. 7.

31 *Ibid.*

32 Hartlib, *Ephemerides*, Sheffield University Library, 1650 FG 8. The 'Elucidarium' was to continue the ideas developed in the *Continuation*. See Turnball, *Hartlib, Dury and Comenius*, pp. 435ff. It was never published.

33 William Petty, *The Advice of W.P. to Mr. Samuel Hartlib, For the Advancement of Some Particular Parts of Learning* (London, 1648) in *Harleian Miscellany* (London, 1810), VI, 141ff.

34 *Ibid.*, p. 144.

35 Petty, 'Exercises of sense and reason', in *Petty Papers*, ed. the Marquis of Lansdowne (London, 1927), no. 83, II, 7.

36 The conjunction of language and science continued. In 1653 his botanical interest

resulted in a treatise on plants which was never published (see *Petty Papers*, II, 261). In 1677 he wrote a taxonomical paper 'On the Scale of Creatures' (*Petty Papers*, no. 87, II, 21ff.). He later proposed a symbolic logic which depended on provision for 'the species of things' ('Of Algebra', *Petty Papers*, no. 86, II, 10ff.). His interest in language continued throughout his life. In 1664 we hear of him discussing the universal character (no doubt Wilkins') with Pepys (8 January 1664, *Diary of Samuel Pepys*, ed. H. B. Wheatley (London, 1893; rpt. New York, 1946), II, 813). See also above, pp. 181–2.

37 Elias Ashmole, trans., *Fasciculus chemicus* by Arthur Dee (London, 1650), Prolegomena.

38 Thomas Urquhart, *Logopandecteision, or an Introduction to the Universal Language* (1653) in *The Works of Sir Thomas Urquhart*, A Maitland Club Edition (Edinburgh, 1834). There is a facsimile edition published by Scolar Press (Menston, 1970). All references are to the Maitland Club Edition. Urquhart had been working on a universal language for some time but his books and papers were lost in the rout of the Loyalist army after the Battle of Worcester, 3 September 1651.

39 Urquhart, *Logopandecteision*, p. 320.

40 *Ibid.*, p. 318.

41 *Ibid.*

42 *The Trissotetras: or A Most Exquisite Table for Resolving all Manner of Triangles* (London, 1645) in Urquhart, *Works*, pp. 71, 79.

43 *Dictionary of National Biography*, ed. Sir Leslie Stephen and Sir Sidney Lee (Oxford 1921–2), XX, 46.

44 In *Ekskubalouron* (*Works*, p. 271) Urquhart refers to Master John Dury, a fellow Scot 'whose learning is highly extolled in Germany and France'. This would seem to refer to Hartlib's colleague. Dury was minister to the English Company of Merchants in Elbing where Hartlib met him in 1628 and he labored for the rest of his life for reunification of Protestant sects. Urquhart himself spent many years abroad (1635–8; 1642–5; *c.* 1655–60). Both Dury and Urquhart were staunch Royalists and this may have brought them together when Dury was in London between 1640 and 1641. Through Dury, Urquhart could have met Hartlib. For Dury, see *Dictionary of National Biography*, VI, 261–3.

45 See above, p. 106.

46 Keith Thomas, *Religion and the Decline of Magic* (Harmondsworth, 1973) passim; Christopher Hill, *The World Turned Upside Down: Radical Ideas During the English Revolution* (Harmondsworth, 1975), passim.

47 John Webster, *Academiarum examen, or the Examination of the Academies* (London, 1654). This has been reprinted in A. Debus, *Science and Education in the Seventeenth Century: the Webster–Ward Debate* (London, 1970). All references are to the original edition.

48 Webster, *Academiarum examen*, p. 24.

49 *Ibid.*, pp. 24–5.

50 *Ibid.*, p. 25.

51 Boehme's writings enjoyed popularity around the middle of the century. Between 1644 and 1663 all of Boehme's manuscripts were published in English. It was probably Boehme, more than any other writer, who gave currency to the idea of

the language of nature. Boehme himself was influenced by Paracelsus. Both shared the idea that the visible world is a manifestation or emenation of the divine; that God encoded his divine knowledge and message in his creation. This connection between the symbolic, material form of creation and the invisible essences led to Paracelsus' doctrine of signatures which we have discussed above.

52 Webster, *Academiarum examen*, p. 26.

53 *Ibid.*, p. 27.

54 *Ibid.*, p. 29. Cf. Boehme's description from *The Threefold Life of Man* in *The Works of Jacob Behmen* (London, 1764), II: '...now as the Spirit of Eternity hath formed and framed all things, so also the spirit of man formeth them in his word, for all ariseth from his Center...he giveth everything its name according to the spirit and form of everything, for the inward speaketh forth the outward' (p. 52). 'Like the spirit in creating things so the human Word goeth also forth out from the Center of the Spirit in shape, property, and form, and it is no other, than the Spirit maketh such a substance, as Creation itself is, when it expresseth the form of Creation' (p. 58). According to Boehme, 'All things visible were emanations of things invisible; the hidden God lay revealed under visible creation. The "outbreathed" or "outspoken" invisible power, the Word, had called forth creation Nature is the language of God to man, if only man will read it aright... He also claimed another special gift, direct insight into the language of nature, the language Adam spoke when he named the animals in the Garden of Eden...The observation was made more than once that the hoped for philosophical language would copy the function of Adam's language, thus recapturing a measure of the insight into nature and unity of knowledge that had been lost' ('Boehme', *Dictionary of Scientific Biography*, ed. C. C. Gillispie (New York, 1970–80), II, 223).

55 Seth Ward, *Vindiciae academiarum* (Oxford, 1654). Reprinted in Debus, *Science*. It is prefaced with a letter written by [Joh]N. [Wilkin]S. I quote from the original edition. In his *Essay*, Wilkins described Webster's tract as a 'torrent of affected insignificant tautologies' ('Epistle to the Reader').

56 Ward, *Vindiciae academiarum*, pp. 20-1.

57 *Ibid.*, p. 21.

58 John Wilkins, *An Essay Towards a Real Character and a Philosophical Language* (London, 1668). There is a facsimile of this edition published by Scolar Press.

59 Ward, *Vindiciae academiarum*, p. 21.

60 Robert Kargon, *Atomism in England from Hariot to Newton* (Oxford, 1966), p. 71. See also E. J. Dijksterhuis, *The Mechanization of the World Picture* (Oxford, 1961), pp. 205–9.

7. Schemes for a philosophical language

1 George Dalgarno, *Ars signorum* (London, 1661). There is a Scolar Press facsimile of the 1st edn (Menston, England, 1968). There is also a reprint in *The Works of George Dalgarno of Aberdeen*, A Maitland Club Publication (Edinburgh, 1834). All quotations are from the Maitland Club edition.

2 Dalgarno to Pell, 12 December 1657, Sloane MS 4377, f. 151r.

3 *Ibid.*

4 Dalgarno, 'News to the Whole World, of the Discovery of an Universal

Character, and a New Rational Language', Sloane MS 4377, f. 143. Reprinted in facsimile in the Scolar Press edition of *Ars signorum* and R. C. A. Alston, *The Bibliography of the English Language from the Invention of Printing to the year 1800* (Leeds, 1965–73), VII, Plate XI.

5 Dalgarno, 'The possibilitie and great usefulnesse of an Universall character', Sloane MS 4377, f. 157. Reprinted in facsimile in Alston, *Bibliography*, VII, Plate XIII.

6 Salmon, 'The evolution of Dalgarno's "Ars signorum"' in *Studies in Language and Literature in Honour of Margaret Schlauch*, ed. M. H. Brahmer, et al. (Warsaw, 1966), p. 366.

7 Oldenburg to Hartlib cited in G. H. Turnbull, 'Samuel Hartlib's influence on the early history of the Royal Society', *Notes and Records of the Royal Society*, X(1953), p. 119.

8 Salmon, 'Evolution', p. 366.

9 *Ibid.* New Dalgarno papers have recently been discovered in the Gregory Collection of Manuscripts at Christchurch, having been given to David Gregory by Dalgarno's widow in 1698. These are being edited by Dr David Cram of the University of Aberdeen. Among the manuscripts is a treatise on universal language in which Dalgarno gives an autobiographical account of the development of his language scheme and also a critique of Wilkins' *Essay*. These should shed more light on the differences between the theories of Dalgarno and Wilkins and on their relationship. See David Cram, 'George Dalgarno on "Ars signorum" and Wilkins' "Essay"', in *Progress in Linguistic Historiography*, ed. Konrad Koerner, Amsterdam Studies in the Theory and History of Linguistic Science: series 3, *Studies in the History of Linguistics*, vol. 20 (Amsterdam, 1980), pp. 113–21.

10 Dalgarno, 'Omnibus omnino hominibus orbis terrarum praecipue dominis', Sloane MS 4377, ff. 139–42. Reprinted in Alston, *The Bibliography of the English Language*, VII, Plate XIV.

11 Robert Plot, *The Natural History of Oxfordshire, being an Essay toward the Natural History of England* (Oxford, 1676), pp. 282–3.

12 *Ars signorum*, p. 37. 'Quare, qui cum Ratione *Nomina Rebus* imponere velit, primo oportet in Chaos istud Mundi Idealis in animo existentis, per quasi creationem Logicam, formam, pulchritudinem, et ordinem introducere...'

13 *Ibid.* '...sic Grammaticus, secundum Ideas et Regulas Logicas a Rerum ipsarum natura extrinsecus existentium petitas, Nomina illis imponere debet. Haec autem ordinata Series Rerum vulgo dici solet Praedicamentum'.

14 *Ibid.* '...Rerum naturas aperiant, differentias et convenientias scrutando, sicque, Methodo et ordine inter se collocando et disponendo'.

15 *Ibid.*, p. 41. '...omnis enim demonstratio supponit definitionem, definitio genus et speciem, genus et species ordinatam seriem Praedicamentalem....nulla est definitio vel demonstratio...quia nulla est constituta series praedicamentalis'.

16 *Ibid.*, p. 43. 'Vel primo, investigando per Analysin Logicam, certum numerum primarum et simplicissimarum Notionum, ex quibus omnes aliae complexae componuntur; et quidem hic Analyticus modus in penitiorem Rerum cognitionem ducit. Novi sententiam esse ingeniosissimorum hominum, numerum Notionum simplicium per strictam Analysin Logicam, pauciorem esse numero

sonorum articulatorum simplicium per strictam Analysin Grammaticam. Vel secundo, constitui potest haec series reducendo omnes Notiones Naturae et Artis, tam Complexas quam Simplices, ad lineam rectam praedicamentalem.'

17 *Ibid.*, pp. 43, 45.

18 'The Tables of the Universal Character', Sloane MS 4377, f. 145.

19 Cram, 'George Dalgarno', p. 118.

20 *Ars signorum*, pp. 52–3. 'Tenendum in Rerum Nominibus componendis, non esse necessarium ut differentia Generi superaddita sit tota rei Forma, quam docent philosophi esse unum aliquid simplex, occultum (ipsi nesciunt quid) latitans invisibiliter (et etiam inintelligibiliter) in Rebus; ad quam inveniendam nullum acumen penetrare potest...omnium enim Rerum quarumcunque formae sunt inadaequate cognitae...Forma enim nihil est aliud quam aggregatum omnium accidentium alicujus Rei.'

21 *Ibid.*, pp. 48–9.

22 *Ibid.*, p. 9.

23 In treating plants Dalgarno indicates his sources are Adrianus Spigelius, *Isagoges in rem herbarium* (Padua, 1606) and Daniel Sennertum, *Epitome universam Dan. Sennerti doctrinam summa fide complectens, ex triplice volumine in unum congestam...* Per Claudium Bonnecium (Piot, 1654). See *Ars signorum*, p. 50. Comparison with the major classes listed in each of these two indicates that Dalgarno took considerable liberty in adding and dropping items.

24 *Ars signorum*, pp. 11–23.

25 *Ibid.*, pp. 12–13.

26 *Ibid.*, p. 12.

27 *Ibid.*, p. 9.

28 *Ibid.*, pp. 11–13.

29 *Ibid.*, p. 9.

30 *Ibid.*, p. 12.

31 *Ibid.*, p. 102.

32 Ralph W. V. Elliott, 'Isaac Newton as phonetician', *Modern Language Review*, 49 (1954), p. 5. The unpublished notebook is in the Pierpont Morgan Library in New York. The source of the arrangement of the word list has been identified as Francis Gregory, *Nomenclatura brevis Anglo-Latin-Graeca*, 4th edn (1654). See Frank E. Manuel, *A Portrait of Isaac Newton* (Cambridge, Ma., 1968). Wilkins had been made Master of Trinity in 1659 but was replaced in 1660. Ray was still in residence at Trinity in 1661.

33 Elliott, 'Isaac Newton's "Of An Universall Language"', *Modern Language Review*, 52 (1957), pp. 1–18. A transcription of the manuscript is provided there. The manuscript is now located at the University of Chicago Library.

34 *Ibid.*, p. 10.

35 *Ibid.*

36 *Ibid.*, p. 12.

37 *Ibid.*

38 *Ibid.*, p. 16.

39 *Ibid.*, p. 15.

40 Comenius, *Panglottia*, in *De rerum humanarum emendatione consultatio catholica* (Prague, 1966), II, 147–204. Included here is 'Novae linguae harmonicae tentamen primum', II, 189–204. Related to these is the *Lexicon reale pansophicum*, probably written in 1662. The *Triertium catholicum*, written just before Comenius' death in 1670, carries on the discussion of language. See V. T. Miškovskà, 'Comenius (Komenský) on lexical symbolism in an artificial language', *Philosophy*, 37 (1962), pp. 238–44. See also John E. Sadler, *J. A. Comenius and the Concept of Universal Education* (London, 1966), pp. 143–63; Miškovskà, 'La Panglottie de J. A. Komenský', *Philologica Pragensia*, II (1959), pp. 97–105; 'Le Triertium et la *Panglottia*', *Acta Comeniana*, 3 (1973), pp. 383–90; 'Langue philosophique et religion au XVIIe siècle', *Communio Viatorum*, 2 (1959), pp. 335–42; 'Empirisme ou invention à la base lexicale de la Panglottie?' *Acta Comeniana*, 20 (1961), pp. 189–97.

41 G. H. Turnbull, *Hartlib, Dury and Comenius*, p. 377. After his departure from England in 1642, Comenius visited Descartes and we know that the subject of universal language was discussed by them (without much interest on the part of Descartes). For the rest of the 1640s and for the first half of the next decade, Comenius was employed first in Sweden and then in Hungary to carry out educational reforms and to write textbooks. After the sack of Leszno in 1656, where he lost all his worldly possessions including his library and manuscripts, Comenius took refuge in Amsterdam where he lived until his death in 1670.

42 Miškovskà, 'Comenius', p. 238. In 1668 Oldenburg writes promising to send him a copy of the *Essay* (5 June, *The Correspondence of Henry Oldenburg*, ed. A. Rupert Hall and Mary Boas Hall (Madison, Wis., 1968–), pp. 449–550).

43 *Panglottia*, pp. 191, 197.

44 *Ibid.*, p. 197.

45 *Ibid.*, p. 202.

8. The culmination, aftermath and end

1 Wilkins, *An Essay Towards a Real Character and a Philosophical language* (London, 1668), To the Reader. All references are to the first edition of 1668.

2 *Ibid.*

3 *Ibid.*

4 *Ibid.*

5 Wallis, *A Defence of the Royal Society and Philosophical Transactions...in Answer to the Cavils of Dr William Holder* (London, 1678), p. 17.

6 Plot, *The Natural History of Oxfordshire being an Essay toward the Natural History of England* (Oxford, 1676), p. 289. In 1657, Walter Charleton, a former pupil of Wilkins, remarked 'how little he wants of finishing that so long talked-of and desired design of an Universal Character and Language' (*Immortality of the Human Soul* (London, 1657), p. 46). In December of that year Boyle wrote Hartlib that 'The Universal Character of Dr. Wilkins had now brought to that perfection that he is pleased to promise me sometime next week to shew it to Dr. Ward and me together, that it may be jointly considered by us all, and if he be as wary in his affirmations about it as he used to be about other things, I make no doubt

but it will be found extraordinarily ingenious, though I confess I yet distrust the practicableness of it in divers cases.' (Quoted by Hartlib in a letter to Pell, 31 December 1657, in Robert Vaughan, ed., *The Protectorate of Oliver Cromwell* (London, 1838), II, 435.) The same month Comenius sent his 'understanding' of it to Dalgarno (which Dalgarno thought was 'not worth imparting'). (Reported in a letter from Hartlib to Boyle, 7 January 1658 in *The Works of the Honourable Robert Boyle*, ed. Thomas Birch (London, 1772), VI, 99.) In 1659, John Owen refers to a discussion he had with Wilkins 'about an universall Character, which hath been mentioned by many, attempted by divers, and by him brought to that perfection, as will doubtlesse yield much, if not universall satisfaction unto learned and prudent men, when he shall be pleased to communicate his thoughts upon it to the world...' (John Owen, *Of the Divine Original* (London, 1659), p. 227).

7 The immediate years following the Restoration saw Wilkins installed in the parish of St Lawrence Jewry with his son-in-law Tillotson whose biographer reports that 'His joining with Dr. Wilkins in perfecting a scheme of a real character and philosophical language...led him to consider exactly the truth of language and style' (Thomas Birch, *The Life of Tillotson* (London, 1752), pp. 21–2).

8 In the same year Thomas Birch reports that 'Dr. Wilkins was put in mind to prosecute his design of an universal language' (Birch, *History of the Royal Society* (London, 1756), p. 119). Wallis expressed his hopes for a universal character which 'must needs be allowed to be very sensible, if not facile'. Wallis himself wanted to see the artificial language expressed in roman letters; he suggested an immediate application of the alphabetic character to teaching language to the deaf. (John Wallis, 'Letter of Dr. John Wallis to Robert Boyle, Esq.; concerning the said Doctor's Essay Teaching a person Dumb and Deaf to speak and to Understand Language', reprinted in *The Philosophical Transactions of the Royal Society*, 18 July 1670, 5, no. 61, p. 1092.) At about the same time there was circulating a popular ballad that can refer to no one but Wilkins: 'A doctor counted very able/Designs that all mankind shall/Spite o' th' Confusion made att Babell/By Character call'd Universall/How long this Character will be learning/That truly passeth my discerning' (Dorothy Stimson, *Science and Amateurs* (New York, 1946), pp. 61–2). In the spring of 1663 Abraham Hill reports that Wilkins 'intends to put his last hand' to the universal character on which he has already taken 'some Pains' (Hill to John Brooke, 19 May 1663 in Abraham Hill, *Familiar Letters* (London, 1767), p. 108). A half year later in January 1664 Pepys reports that he and Petty 'fell to talk of musique, the universal character, arte of memory, etc.' (8 January 1664, Samuel Pepys, *The Diary of Samuel Pepys*, ed H. B. Wheatley, I, 813). The following year, 1665, was interrupted by the Great Plague which struck London, during which time Wilkins withdrew from the city with Petty and Hooke.

9 11 January 1666, Pepys, *Diary*, II, pp. 132–3.

10 Sprat, *History of the Royal Society* (London, 1667), ed. Jackson I. Cope and Harold Whitmore Jones (St Louis, Mo., 1958), p. 251.

11 Hooke, 'A General Scheme or Idea of the present state of Natural Philosophy and how its defects may be remedied by a Methodical Proceeding in the Making of Experiments and collecting Observations, whereby to compile a Natural History

as the solid basis for the Superstructure of Philosophy', in *The Posthumous Works of Robert Hooke*, ed. Richard S. Westfall (New York, 1969). This is a reprint of the original edition by Richard Waller published in London in 1705. On Hooke's scheme see Mary B. Hesse, 'Hooke's philosophical algebra', *Isis*, 57 (1966), pp. 67–83.

12 Hooke, 'General Scheme', p. 21.

13 *Ibid.*, pp. 23–5.

14 *Ibid.*, p. 26.

15 *Ibid.*, p. 64.

16 Wilkins to Willoughby, 20 October 1666, in Ray, *Philosophical Letters Between the late learned Mr. Ray and Several of His ingenious Correspondents*, ed. William Derham (London, 1718), p. 366. See De Mott, 'Science versus mnemonics', *Isis*, 48 (1957), pp. 3–12.

17 Ray, *Philosophical Letters*, ed. Derham, p. 366.

18 *Ibid.* Since Ray and Wilkins were colleagues at Trinity College in 1660 it is possible that at that time Ray knew of the language schemes of Wilkins and Ward, as well as of the schemes of Dalgarno and Newton. Ray stayed at Trinity after the Restoration but in 1662 resigned his Fellowship rather than take the Oath of Uniformity. He worked for a while on an appendix to his catalogue of Cambridge plants (1663), then went to Europe with Willoughby. It was presumably during his European sojourn that he converted to the philosophical systematics of Cesalpino. Ray and Willoughby returned to England in the spring of 1666. In November 1667 Ray was elected to the Royal Society.

19 Ray to Lister, 18 June 1667, in Ray, *Further Correspondence*, ed. R. W. T. Gunther (London, 1928), p. 112. At this time, Wilkins had nothing if not grandiose plans: 'If I could fully satisfy myself in the methodical enumeration I would put out the next edition in Folio with handsome cuts of all such things as are fit to be represented in Figure' (Ray, *Philosophical Letters*, ed. Derham, pp. 366–7).

20 4 June 1966, Pepys, *Diary*, II, 210.

21 H. C. Foxcroft, *A Supplement to Burnet's History of My Own Time* (Oxford, 1902), p. 487; Gilbert Burnet, *A History of My Own Time* (London, 1833), I, 345. Lloyd and Wilkins travelled in the same circles. Between 1656 and 1659 Lloyd was a private tutor at Wadham. Forced to leave the University with the change in government, in 1666 he was appointed one of the King's Chaplains (as was Wilkins). In 1672 he was installed as Dean of Bangor and in 1680 consecrated as Bishop of St Asaph. In the later years of his life (he lived to be 90) Lloyd was 'half crazed by an unremitting study of the apocalyptic visions' and 'came to number himself among the prophets' (*Dictionary of National Biography*, XI, 1316).

22 On Lodowyck, see Wilkins, *Essay*, p. 357; on Buckingham see Bodleian Library Aubrey MS 13, f. 39v.

23 29 November 1667, Pepys, *Diary*, II, 727.

24 Thomas Birch, *The History of the Royal Society* (London, 1756), II, 265.

25 *Ibid.*, p. 281.

26 15 May 1668, Pepys, *Diary*, II, 873.

27 John Collins to Pell, 28 May 1668, Sloane MS 4278 cited by Shapiro, *John Wilkins: An Intellectual Biography 1614–1672* (Berkeley, Ca., 1969), pp. 220, 313n.

28 Wilkins, *Essay*, p. 20.
29 *Ibid.*
30 *Ibid.*, p. 21.
31 *Ibid.*, p. 22.
32 *Ibid.*, p. 289.
33 *Ibid.*
34 *Ibid.*, p. 20.
35 *Ibid.*, p. 23.
36 *Ibid.*, p. 69.
37 Ray, *Methodus plantarum nova* (London, 1682), p. 3.
38 Wilkins, *Essay*, p. 67.
39 *Ibid.*, p. 68.
40 *Ibid.*, p. 69.
41 *Ibid.*, p. 70.
42 *Ibid.*, pp. 81, 88. In the tables for the three kinds of herbs, Ray found it impossible to limit the number of 'species' to 9 so Wilkins devised a way to extend the number to 15: 'As for those Species under Plants and Animals, which do exceed the number of Nine, they may be expressed by adding the letters L or R after the first consonant, to denote the second or third of such combinations. Thus if *Gade* be Tulip, viz the Third Species in the first Nine, then *Glade* must signify Ramson, the third in the second Nine, or the 12th Species under that Difference' (*Essay*, p. 415). In the taxonomical tables, each of the species is arranged in pairs. The tables spell out the features by which each member is distinguished from the other but there are no letters in the designatory names that signify these final differences.
43 *Ibid.*, p. 318.
44 *Ibid.*
45 Cram, 'George Dalgarno on "Ars signorum" and Wilkins' "Essay"', *Progress in Linguistic Historiography*, ed. Konrad Koerner (Amsterdam, 1980), p. 119.
46 *Ibid.*
47 Birch, *History*, II, 283.
48 *Ibid.* In October, 1668 'the President moved that the persons formerly desired to peruse and consider the Bishop of Chester's *Essay*...should be called up at the next meeting to bring in their thoughts concerning that book' (Birch, *History*, II, 315). Evidently there was not sufficient interest, for the *Essay* is never mentioned again.
49 Wallis, *Defence*, p. 17. These letters are preserved in the Bodleian Library, Savile MS 4.
50 Beck to Oldenburg, 15 August 1668, in Royal Society Letter Book, B1, f. 137.
51 John Webb, *An historical Essay endeavoring a Probability that the Language of the Empire of China is the Primitive Language* (London, 1669), p. 187.
52 Simpson was prompted to air the ideas he had been considering on the subject of a universal character, viz. that various wise men should construct a character based either on hieroglyphics, numbers or musical tones. The language itself should be composed of 'so many Roots or Radical Characters as might express the most known and useful things in the world; as, Man, Book, House, Stone,

etc. which should be orderly placed down in a vocabulary' (William Simpson, *Hydrologia chymica* (London, 1669), p. 224).

53 Robert Hooke, *A Description of Helioscopes and some other instruments* (London, 1676), pp. 30–1.

54 The figure is found in Margaret 'Espinasse, *Robert Hooke* (London, 1956), p. 72. C. N. de C. Andrade, 'The Real Character of Bishop Wilkins', *Annals of Science*, I (1936), p. 12 translates it as follows: 'The contriving of the portable watches is comprehended as far as their regulation is concerned by two objects. The first is the causing of a counter effect of all gravity and motion, so as neither the pulling power of the earth nor position nor irregular external motion of the watch can cause any irregularity of this. The second is the contriving of internal regulators of the watch so as neither the irregular functioning power of the spring of the bows [mainspring?] or other part of the watch can disturb its motion. The first is effected by balances moving continually and oppositely, the first balance to the other. The second is effected by springs or other agents acting independently of the balances or regulators of the motion.'

55 Plot, *The Natural History of Oxfordshire, being an Essay toward the Natural History of England* (Oxford, 1676), p. 290.

56 *Histoire des Séverambes* (Paris, 1678–9), Partie II. The first part had originally been published in London. Locke was also part of this group and is known to have owned a copy. See Knowlson, 'The ideal languages of Veiras, Foigny, and Tyssot de Patot', *Journal of History of Ideas*, 24 (1963). Gabriel Foigny, *La Terre australe connue* (Paris, 1676) also included an artificial language, the source for which was Comenius' *Via lucis*. Tyssot de Patot's language in *Voyages et avantures de Jacques Masse* (Bordeux, 1710) is traced by Knowlson to other voyages and to Father Philippe Labbé's *Grammatica linguae universalis* (1663). See Paul Cornelius, *Language in 17th and 18th Century Imaginary Voyages* (Geneva, 1965), passim; Seeber, 'Ideal languages in the French and imaginary voyages', *Publications of the Modern Language Association*, LX (1945), pp. 586ff. and Emile Pons, 'Les Langues imaginaires dans le voyage utopique: les deux grammairiens: Vairasse et Foigny', *Revue de Littérature Comparée*, 12 (1932), pp. 500–32.

57 Neremiah Grew, *Musaeum regalis societatis or a Catalogue and Description of the Natural and Artificial Rarities belonging to the Royal Society and preserved at Gresham College* (London, 1681) cited by Dorothy Stimson, *Science and Amateurs* (New York, 1946), p. 111.

58 Wallis to Borelli, 11 January 1670, in Stephen Rigaud, *Correspondence of Scientific Men of the Seventeenth Century* (Oxford, 1841), II, 518.

59 Walter Pope to Aubrey, 24 May 1677, Aubrey MS 13. Newton, *Correspondence*, ed. H. W. Turnbull (Cambridge, 1959–), II, 296.

60 Huygens to Moray, 30 March 1669, in Christian Huygens, *Oeuvres completès*, Société Hollandaise des Sciences (The Hague, 1888–1950), VI, 397.

61 Moray to Huygens, 26 April 1669, in Huygens, *Oeuvres*, VI, 425.

62 Toinard to Locke, 6 July 1681 in Locke, *The Correspondence of John Locke*, ed. E. S. De Beer (Oxford, 1976–), II, 422. Locke to Toinard, 30 August 1681, *Correspondence*, II, 436.

63 William Wotton, *Reflections on Ancient and Modern Learning* (London, 1694), p. 60.

64 *The Athenian Oracle, a Collection of Mercuries* (London, 1704), I, 335–6.

65 Phillip Skippon to Ray, 13 December 1667, in Ray, *The Correspondence of John Ray*, ed. Edwin Lankester (London, 1848), p. 22.

66 Ray to Martin Lister, n.d., in Ray, *Correspondence*, ed. Lankester, p. 55.

67 Mark Lewis, *An Apologie for Grammar* (London, 1671), p. 39.

68 Robert Hooke, *The Diary of Robert Hooke*, ed. Henry Robinson and Walter Adams (London, 1935), pp. 203-4.

69 Ray to Aubrey, 3 March 1677, in John Ray, *Further Correspondence*, ed. R. W. T. Gunther, p. 163.

70 William Derham, 'The life of John Ray', in Ray, *Memorials of John Ray*, ed. Edwin Lankester (London, 1846), p. 22.

71 Ray to Aubrey, 8 May 1678, in Ray, *Correspondence*, ed. Lankester, p. 128. Lewis du Moulin was Professor of History at Oxford. His brother, Peter, was tutor to the Boyle family during the Civil War and then tutor at Oxford.

72 Evelyn to Pepys, 2 September 1694, in Samuel Pepys, *Letters and the Second Diary of Samuel Pepys*, ed. R. G. Howarth (London, 1932), pp. 248–9.

73 Ray to Lister, 28 April 1670, in *Correspondence*, ed. Lankester, p. 55.

74 Ray, *Catalogus plantarum Angliae*, 2nd edn (London, 1677) with 'Errata', n.p.

75 John Aubrey, *Brief Lives*, ed. Oliver Dick (London, 1958), p. 320.

76 Paschall to Glanvill, 10 June 1669, in Royal Society Letter Book, P 1, f. 53.

77 Andrew Paschall, *Propositions for Carrying on a Philosophical Correspondence, already begun in the County of Somerset upon incouragement given from the Royal Society* (London, 1670), cited by Vivian Salmon, 'John Wilkins' *Essay* (1668): critics and continuators', *Historiographia Linguistica*, I (1974), p. 150. The Wing Catalogue lists this as P3786. According to Wing, there are copies at the British Museum and at the Bodleian Library. Neither library was able to locate the treatise for me.

78 Paschall to Aubrey, 12 June 1672, in Aubrey MS 13, f. 2.

79 Lodowyck, 'An Essay Towards an Universal Alphabet' and 'A Second Essay Concerning a Universal Primer'. Originally in pamphlet form in 1685 they were then reprinted in the *Philosophical Transactions of the Royal Society*, 16, no. 182 (1686), pp. 126–37. This is a later version of the papers in the Sloane manuscripts and Royal Society Classified Papers: 'Charracters for the new alfabet', Sloane MS 897, ff. 24v-25r; 'Essay of an universall perfect alfabet', Sloane MS 897, ff. 25v–31v; 'Concerning a perfect, universall alfabeth', Sloane MS 932, ff. 1r–3v; 'Concerning the caracters for this universalphabet [sic], Sloane MS 932, f. 4; 'A designe towards an universall alfabet', Royal Society Classified Papers, 16, nos. 4, 5; 'That the alfabet and orthografy of the English tongue is defective...' Sloane MS 1466, ff. 214r–216r. See Lodowyck, *The Works of Francis Lodwick: A Study of his writings in the intellectual context of the 17th century*, ed. Vivian Salmon (London, 1972), pp. 152ff. and p. 160. It would be a misnomer to call this universal alphabet a universal language for it is no more than a system for phonetic transcription embodied in a universal written character, this on the grounds that it would be easier to establish a new alphabet than to remedy existing ones ('Second essay', p. 137).

80 Paschall to Aubrey, 16 March 1674, Aubrey MS 13, f. 3.

81 Paschall to Aubrey, 13 January, 1674, Royal Society Letter Book, P 1, f. 55.

82 Paschall to Aubrey, 16 October 1676, Aubrey MS 13, f. 12.
83 Paschall to Aubrey, 8 July 1678, Aubrey MS 13, ff. 32–3. See A. J. Turner, 'Andrew Paschall's Tables of Plants for the Universal Language (1678)', *Bodleian Library Record*, LX (1978), pp. 346–50.
84 Aubrey MS 13, ff. 32–3.
85 Aubrey to Ray, 9 July 1678, Ray, *Correspondence* ed. Lankester, p. 129. See also *Aubrey on Education*, ed. J. E. Stephens (London, 1972), p. 35.
86 Hooke, *Diary*, passim. See esp. 9 December 1675, p. 199 and 18 December 1675, p. 202.
87 Pigott to Aubrey, 15 October 1676, Aubrey MS 13, f. 105. According to Aubrey, Pigott was born in 1657, received an M.A. at Oxford, was a fellow at Wadham College and Chaplain to the Earl of Ossory. He died in 1686 (Aubrey, *Brief Lives*, ed. Andrew Clark (Oxford, 1898) II, 155).
88 Salmon, 'John Wilkins' *Essay* (1668): critics and continuators', *Historiographia Linguistica*, I (1974), p. 155. See Sloane MS 859, f. 3. Paschall was also aware of the inadequacies of the tables, particularly in that they omit words for Sciences, Arts and Trades (which Hooke later worked on). He worried also about the classifications, some of which were incorrect. He decided however to hold off on his revisions of the tables until he had a clearer notion of Ward's scheme, which despite many requests was still not forthcoming (Paschall to Aubrey, 26 March, Aubrey MS 13, f. 16v.).
89 Paschall to Aubrey, 13 February 1677, Aubrey MS 13, f. 15.
90 21 December 1677, Hooke, *Diary*, p. 335.
91 8 January 1678, Hooke, *Diary*, p. 339.
92 15 April 1678, Hooke, *Diary*, p. 353.
93 Petty to Aubrey, 29 May 1678, Aubrey MS 13, f. 100.
94 John Aubrey, *The Idea of Education of Young Gentlemen* (1678) in *Aubrey on Education*, ed. Stephens.
95 *Ibid.*, pp. 56, 118.
96 *Ibid.*, p. 126.
97 11 January 1679, Hooke, *Diary*, p. 393.
98 See 20 April 1679, 14 June 1679 and 24 June 1679, Aubrey MS 13, ff. 120, 122, 124.
99 Halley to Aubrey, 16 November 1679, Aubrey MS 13, f. 148.
100 Paschall to Aubrey, 10 November 1676, Aubrey MS 13, f. 13.
101 Pigott to Aubrey, 25 February 1677, Aubrey MS 13, f. 16r.
102 *Ibid.*, f. 17r.
103 *Ibid.*, f. 16.
104 *Ibid.*
105 *Ibid.*
106 *Ibid.*
107 Pigott to Aubrey, 23 February 1678, Aubrey MS 13, f. 110.
108 30 March 1678, Hooke, *Diary*, p. 351.
109 Pigott to Aubrey, 15 April 1678, Aubrey MS 13, ff. 113–14. See also James Knowlson, *Universal Language Schemes in England and France 1600–1800* (Toronto, 1975), pp. 105ff.

110 Paschall to Aubrey, 28 May 1678, Aubrey MS 13, f. 30. George Ritschel, *Contemplationes metaphysicae* (Oxford, 1648). Ritschel was born in Bohemia and attended Oxford c. 1641. In 1644 he was hired as assistant to Comenius in order to help him with the metaphysical portion of his *Janua Rerum.* From 1645 to 1646 he was working on this project at the Bodleian and was in close contact with Hartlib. Comenius and Ritschel parted ways in 1647 in part because their metaphysical notions were quite different and in part because Ritschel objected to the attention Comenius and Hartlib were paying to Kinner who had succeeded him as Comenius' assistant. In 1648 Ritschel published his book at Oxford although because of its Aristotelian bent it attracted little attention in England. In the same year he was appointed as headmaster to a grammar school at Newcastle-upon-Tyne and in the 1650s he moved to Hexham where he remained until his death in 1683. In 1671 Leibniz became interested in Ritschel's book for its bearing on his *ars combinatoria.* See Robert Young, *A Bohemian Philosopher at Oxford in the 17th Century* (London, 1925).

111 Aubrey MS 13, f. 30.

112 *Ibid.*

113 See Pigott to Aubrey, 21 May 1678, Aubrey MS 13, f. 115.

114 Paschall to Hooke, 21 February 1680, Royal Society Letter Book, P 1, f. 57.

115 *Ibid.*

116 *Ibid.*

117 Paschall to Aubrey, 6 February 1680, Aubrey MS 13, f. 38. Hooke replied directly to Paschall that he also feared that Ward's scheme of 100 transcendental notions was unworkable and that Paschall's comparison 'of their two ways to the two ways of Arithmetic does very much clear the notion'. (Hooke to Paschall, 11 March 1680, Royal Society Letter Book, H 3, f. 61.) But, he fears, Wilkins' scheme is not 'brought to the highest perfection'. (*Ibid.*)

118 Paschall to Aubrey, 20 March 1680, Aubrey MS 13, f. 39.

119 *Ibid.*

120 *Ibid.*

121 Paschall to Aubrey, 13 November 1682, Aubrey MS 13, ff. 56-7.

122 Paschall to Aubrey, 6 February 1680, Aubrey MS 13, f. 38.

123 Paschall to Aubrey, 2 March 1680, Aubrey MS 13, f. 39.

124 Pigott to Aubrey, 19 February 1680, Aubrey MS 13, f. 126.

125 Paschall to Aubrey, 2 March 1680, Aubrey MS 13, f. 39.

126 Paschall to Aubrey, 10 November 1680, Aubrey MS 13, f. 43.

127 Paschall to Hooke, 21 October 1681, Royal Society Letter Book, P 1, f. 59.

128 Paschall to Lodowyck, 9 August, 1682, Royal Society Letter Book, P 1, f. 60.

129 Paschall to Aubrey, 13 November 1682, Aubrey MS 13, ff. 56-7.

130 Pigott to Aubrey, c. 1683, Aubrey MS 13, f. 135.

131 The silence of Pigott and Paschall may be the result of Ward's condition. According to the *Dictionary of National Biography*, ed. Sir Leslie Stepben and Sir Sidney Lee (Oxford, 1921-2), 'In his later years, Ward's intellect became much weakened' (XX, 796). He was involved in a controversy with his dean and although in 1683 he was victorious, 'when the excitement of the controversy had passed, he sank into complete senility' (*ibid.*). Pigott died suddenly in 1686.

132 R. W. T. Gunther, *Early Science in Oxford* (Oxford 1928), XII, 181; *Petty Papers*, ed. The Marquis of Lansdowne (London, 1927), II, 87.
133 Petty to Southwell, 24 November 1687, in *Petty Papers*, II, 150. The 'Dictionary' and 'Twelve Theological Words' are reprinted in the same volume.
134 Hooke, *Diary*, passim.
135 For a full description of Leibniz' correspondence on the universal language see C. I. Gerhardt, *Die Philosophischen Schriften von Gottfried Wilhelm Leibniz* (Berlin, 1890), VII, 4–25; see also Oldenburg, *Correspondence*, VII, 67–8; VIII, 28–9; IX, 597–601.

For a description and discussion of Leibniz' interest in a real character and universal language (grammar) as well as the relation of these to his plans for an art of combinations and universal science see Louis Couturat, *La Logique de Leibniz* (1901; rpt. Hildesheim, 1961). When Leibniz visited the Royal Society in 1673 he had told Boyle and Oldenburg 'that he did not think either Dr. Wilkins or Mr. Dalgarno had come to the point...they had not obtained the real character, which would be the best instrument of the human mind, and would extremely assist both the reason and memory, and the invention of things'. Leibniz argued that 'These characters ought to resemble as much as possible those of algebra, which are very simple and expressive, and are never superfluous or equivocal, but whose varieties are grounded on reason' (*Éloge de M. Leibnitz*, par M. de Fontenelle, in Dugald Stewart, *The Collected Works of Dugald Stewart*, ed. Sir William Hamilton (Edinburgh, 1854), II, 197). In a letter to Burnet in 1697 Leibniz said of the *Essay* that there had never been a more accomplished table of predicates. But he repeated his charge that this character did not do what was necessary to be done, i.e. correspond to the analysis of thoughts. (Leibniz to Burnet, 24 August 1697, Leibniz, *Die Philosophischen Schriften*, ed. C. I. Gerhardt, III, 216).
136 Hooke to Leibniz, 12 July 1680, Royal Society Letter Book, H 3, f. 63. See also Hooke to Leibniz, 19 May 1681, Royal Society Letter Book, H 3, f. 64.
137 Hooke to Leibniz, 12 July 1680, Royal Society Letter Book, H 3, f. 63.
138 See Hesse, 'Hooke's philosophical algebra'.
139 Hooke, 'On Memory', in *The Posthumous Works of Robert Hooke* (London, 1705), ed. Richard S. Westfall (New York, 1969), p. 138.
140 Hooke, 'Mathematical Language', Royal Society Classified Papers, 20, no. 72.
141 *Ibid.*
142 Ray to Lister, 7 May 1669, in Ray, *Correspondence*, ed. Lankester, pp. 41–2. Translated from the Latin in De Mott, 'Science', p. 5.
143 Ray to Lister, 28 April 1670, in Ray, *Correspondence*, ed. Lankester, pp. 55–6.
144 Ray to Aubrey, 6 May 1678, in Ray, *Further Correspondence*, ed. Gunther, pp. 163–4.
145 *Ibid.*, p. 164.

9. The fall of essentialist taxonomy

1 Thomas Baker, *Reflections Upon Learning* (London, 1700), pp. 20–1.
2 Robert Kargon, *Atomism in England from Hariot to Newton* (Oxford 1966), p. 107.
3 *Ibid.*, pp. 109ff.

4 The views Wilkins expressed in *Of the Principles and Duties of Natural Religion* (London, 1675) were not new at this time; similar ideas were expressed in sermons dating to 1660–2. See Henry Van Leeuwen, *The Problem of Certainty in English Thought 1630–1690* (The Hague, 1963), p. 58.
5 Wilkins, *Natural Religion*, p. 5.
6 *Ibid.*, p. 9.
7 *Ibid.*, p. 5.
8 *Ibid.*, p. 7.
9 *Ibid.*, p. 56.
10 *Ibid.*, p. 5.
11 *Ibid.*, pp. 56–7.
12 A. Rupert Hall, *The Scientific Revolution* (1954; rpt. London, 1967), p. 207.
13 Boyle, as we know, was associated with Wilkins and the Royal Society group from its early days in Oxford. As a central figure on the scientific scene, his early associations place him squarely among the various figures – Hartlib, Pell, Petty, Ward, Wilkins, Dalgarno, Hooke, Locke and Newton – connected with the schemes for the development of a universal language. He was an admitted devotee of 'his master' Bacon.
14 Kargon, *Atomism*, p. 100.
15 Boyle, *The Origin of Forms and Qualities According to the Corpuscular Philosophy* (1666) in *The Works of the Honourable Robert Boyle*, ed. Thomas Birch (London, 1772), III, 42.
16 Kargon, *Atomism*, p. 100. See Boyle, *Origin of Forms*, *Works*, III, 29.
17 *Origin of Forms*, *Works*, III, 42.
18 Boyle, *Experiments, Notes, etc. about the Mechanical Origin or Production of Divers Particular Qualities* (1675), *Works*, IV, 232.
19 *Origin of Forms*, *Works*, III, 27.
20 *Ibid.*, III, 49–50. See Hans Aarsleff, 'Leibniz on Locke on language', *American Philosophical Quarterly*, I (1964), p. 177.
21 *Origin of Forms*, *Works*, III, 12, 15, and passim.
22 *Ibid.*, III, 9.
23 *Ibid.*, III, 41.
24 Boyle, *The Excellency and Grounds of the Corpuscular or Mechanical Philosophy* (1674), *Works*, IV, 78.
25 Locke matriculated at Oxford in 1652 at the age of 20, took a B.A. in 1656 and an M.A. in 1658. At this time Wilkins was conducting informal scientific meetings at Wadham College, Seth Ward was Savilian professor of astronomy; Locke's schoolmate, Hooke, was assisting Willis. At Wilkins' instigation, Boyle had come to Oxford in 1654. In 1660 Locke was appointed as a tutor in classics at Christ Church. Locke was acquainted with Wilkins' group, particularly with Boyle. It is probable that among his teachers were Ward and Wilkins and their friends Wallis and Lewis Du Moulin. See Henry Richard Fox-Bourne, *The Life of John Locke* (London, 1876), pp. 55, 309. Lewis Du Moulin was engaged by Aubrey to translate the real character into French. See also Leeuwen, *Problem*, p. 122. Locke probably joined the experimental group in 1659 through the offices of his school friend, Richard Lower, then assistant to the physician-professor, Thomas

Willis. See Maurice Cranston, *John Locke, A Biography* (New York, 1957), pp. 39ff.; Kenneth Dewhurst, *John Locke (1632–1704), Physician and Philosopher: A Medical Biography* (London, 1963), p. 3. In a letter dated 1660 Locke expressed a desire to make further acquaintance with Boyle who had just returned to Oxford after a two year absence (Locke to Dr Ivie, 20 May 1660, in Dewhurst, *John Locke*, p. 7). By 1663 Locke was informally studying chemistry with Boyle and Boyle's protégé, Peter Stahl (Dewhurst, *John Locke*, p. 10). By 1664 he was closely associated with Boyle and Hooke. In 1665, after returning from a diplomatic mission to Brandenburg as secretary to Sir Walter Vane, Locke and Boyle were in Oxford waiting out the London plague. Boyle was at this time in close correspondence with Wilkins. Locke's cooperation with Boyle was at its height when Boyle published *The Origin of Forms and Qualities* (1666), the most fully developed statement of his corpuscular philosophy (Dewhurst, *John Locke*, pp. 17–19).

26 Barbara Shapiro, *John Wilkins: An Intellectual Biography 1614–1672* (Berkeley, Ca., 1969), p. 202. Locke stayed at Oxford until the spring of 1667 when he moved to London as physician to Anthony Ashley Cooper, First Earl of Shaftesbury. In 1668 he was elected a Fellow of the Royal Society but, although he followed many of their activities with interest, he did not regularly attend the Society's meetings. Wilkins at this time was presiding as Secretary of the Society; Boyle and Hooke were also residing in London and active in the Society.

27 Fox-Bourne, *Life of John Locke*, I, 309.

28 Cranston, *John Locke*, p. 114. See also Shapiro, *John Wilkins*, pp. 154–5.

29 It may be inconsequential but it is an interesting sidelight that around 1664 Locke learned shorthand, presumably in preparation for his diplomatic mission with Vane. Throughout his life, Locke wrote in shorthand and Von Leyden has identified the majority of these writings to be in Jeremy Rich's system. His adoption of Rich's system dates from 1664–6. Von Leyden calls attention moreover to another shorthand alphabet entered by Locke into one of his early account books (*c.* 1661–6) which Von Leyden has not been able to identify. Also there are three notebooks from 1667 which 'contain lists of alternative signs for one and the same word'. (John Locke, *Essays on the Law of Nature*, ed. W. Von Leyden (Oxford, 1958), pp. 248–9). It would be interesting to determine if this interest in shorthand systems had any relation to Wilkins' universal language project or to Dalgarno's *Ars signorum* or to the projectors' general interest in alternative writing systems.

30 The only direct mention Locke makes of the *Essay* is in a letter to his French friend, Nicholas Toinard in 1681. Toinard had asked Locke to send a copy of Dalgarno's *Ars signorum* and Locke replied that Wilkins' *Essay* might be more suitable. (Locke, *The Correspondence of John Locke*, ed. E. S. De Beer (Oxford, 1976), II, 422, 436). Locke also is known to have possessed a copy of Veiras' *Histoire des Sévarambes* which contains an imaginary language based on Wilkins' language scheme.

31 John Locke, *An Essay Concerning Human Understanding*, ed. A. C. Fraser (Oxford, 1894), 3.11.2. All references are to this edition.

32 Dewhurst, *John Locke*, pp. 28–32. Sydenham had also been a pupil of Wilkins.

33 In his undergraduate days Locke is known to have studied plants and to have catalogued 1600 specimens (Dewhurst, *John Locke*, p. 8). Locke's library as well as

his journal notes give us some indication of his biological knowledge: he possessed seventy-four books on botany and natural history which include the works of the standard taxonomers. His letters and journals contain many passing references to his interest in botany and taxonomy. (See J. Harrison and P. Laslett, *The Library of John Locke*, 2nd edn (Oxford, 1971), p. 18; *Locke's Travels in France*, ed. John Lough (Cambridge, 1965), pp. 242–3; Locke, *Correspondence*, passim.)

34 See Locke, *An Early Draft of Locke's Essay together with excerpts from his Journals*, ed. R. I. Aaron and J. Gibb (Oxford, 1936), pp. xii–xxvii for the history of the manuscript.

35 Locke, *Essay*, 2.21.75. As Maurice Mandelbaum, *Philosophy, Science and Perception* (Baltimore, Md, 1964), pp. 4–5, points out, this acceptance of corpuscular philosophy is fundamental to Locke and dates back at least to Draft B of 1671.

36 See *Essay* 2.8.9–10; 2.8.17–18; 4.3.11. See also Richard I. Aaron, *John Locke*, 3rd edn (Oxford, 1971), pp. 117–19.

37 Aaron, *John Locke*, p. 108. See *Essay* 2.8.13; 4.2.11–13.

38 *Ibid.*, p. 111. See *Essay* 2.2.1; 2.1.20–5.

39 *Essay*, 2.23.1.

40 *Ibid.*, 3.3.15.

41 *Ibid.*, 3.3.13.

42 *Ibid.*

43 Aaron, *John Locke*, pp. 203–4.

44 See *Essay*, 3.3.13–17; 2.23.11; 4.3.23–4.

45 *Essay*, 3.6.13.

46 *Ibid.* 3.6.30.

47 *Ibid.*, 3.6.12. See Philip Sloane, 'John Locke, John Ray, and the problem of natural system', *Journal of the History of Biology*, 5 (1972), pp. 1–53. See also the journal entries for 'species', September 1676 and November 1677 in Locke, *An Early Draft*, ed. Aaron and Gibb, pp. 83 and 98–9.

48 *Essay*, 3.6.16, 17, 23, 26, 27.

49 John Yolton, *Locke and the Compass of Human Understanding* (Cambridge, 1970), p. 45.

50 *Essay*, 4.3.23–4. See Yolton, *Locke and the Compass*, pp. 45ff.

51 *Essay*, 4.3.25–6.

52 Patrick Romanell, 'Locke and Sydenham: A Fragment on Smallpox (1670)', *Bulletin of the History of Medicine*, XXXII (1958), p. 316. See also David Givner, 'Scientific preconceptions in Locke's philosophy of nature', *Journal of the History of Ideas*, 23 (1962), pp. 340–54.

53 *Essay*, 3.9.16.

54 *Ibid.*, 3.3.1–5.

55 *Ibid.*, 3.3.20.

56 Aaron, *John Locke*, pp. 203ff. There is considerable debate as to whether or not Locke's view was in fact one of naïve representationalism as this implies. See Norman Kretzmann, 'The main thesis of Locke's semantic theory', *Philosophical Review*, 77 (1968), pp. 175–96; Yolton, *John Locke and the Compass*, pp. 208ff.; Stephen K. Land, *From Signs to Propositions* (London, 1974), pp. 10ff. We are

concerned only with Locke's views as they were interpreted by his contemporaries; what they understood his philosophy to be was 'naïvely representational'.

57 *Essay*, 3.11.24. See Yolton, *Locke and the Compass*, pp. 206–7.

58 *Essay*, 3.11.24.

59 *Ibid.*, 3.11.35. See Yolton, *John Locke and the Compass*, p. 73.

60 John Locke, *Conduct of the Understanding* (1706) in *The Educational Writings of John Locke*, ed. John W. Adamson, 2nd edn (Cambridge, 1922), p. 233. See also *The Educational Writings of John Locke*, ed. James L. Axtell (Cambridge, 1968) and *John Locke and Education*, ed. John W. Yolton (New York, 1971).

61 John Locke, *Some Thoughts Concerning Education* (1693) in *Educational Writings*, ed. Adamson, p. 138.

62 *Essay*, 3.11.2.

63 Ray, *Further Correspondence of John Ray*, ed. R. W. T. Gunther (London, 1928), p. 112.

64 See Robinson to Ray, 9 May 1685, in Ray, *Correspondence*, ed. Lankester, p. 167. See also Ray's reply: 'But that there are fixed and physically indivisible principles in nature I thus argue: If there be no such, but bodies are infinitely divisible, how can there be any constancy in generations or productions? Why are there not infinite new concrete and mixed bodies daily produced, and as many lost?' (12 May 1685, p. 168).

65 John Ray, *The Wisdom of God Manifested in the Works of Creation* (London, 1691), pp. 47–8. These were originally sermons given at Trinity College, Cambridge when Ray was a Fellow.

66 Between 1668 when Wilkins was appointed Bishop of Chester and 1672 when he died, Ray was a frequent visitor, amending the tables and working on the Latin translation. When he was not at Chester, Ray was with Willoughby at Middleton and Wilkins often stopped there. Despite the amount of time that Ray spent with Wilkins his attitude was ambivalent – whether to the man or to the project is hard to determine. In 1670 he writes to Lister that he wants to get Wilkins' project over and done with, but the relationship continued until Wilkins' death. (See Ray to Lister, 28 April 1670, Ray, *Correspondence*, ed. Lankester, pp. 55–6.)

67 After publication of the *Essay*, Ray published one botanical tract, *Catalogus plantarum Angliae* (1670). It was no doubt due to his association with Wilkins that he also published several linguistic works. One was *A Collection of English Proverbs* (London, 1670). Another is *Dictionariolum trilingue* (London, 1675). The fourth edition, entitled *Nomenclator Classicus sive Dictionariolum trilingue* (London, 1703), shows a nice conjunction of the taxonomic and linguistic: 'having lately had occasion to review some of the last published English and Latin nomenclatures, I observed in them some inveterate Errors...That which I chiefly minded was to correct such manifest mistakes as the little insight I have into the History of Plants and Animals enabled me to discern in former nomenclatures...' In the *Nomenclator*, words are arranged under thirty-three headings which in effect make up a taxonomical scheme of all things and which bear clear relations with the philosophical tables. In 1674 Ray published *A Collection of Words not Generally Used...with Catalogues of English Birds and Fishes; and an Account of the preparing*

and referring such Metals and Minerals as are gotten in England. This was the first attempt at a dialect dictionary in English.

68 Ray, *Catalogus plantarum Angliae*, A4.

69 Ray, *Historia plantarum* (1686), I, 40.

70 Ray to Dr Hans Sloane, 11 February 1684, in *Correspondence*, ed. Lankester, pp. 160–1.

71 Ray, *Historia plantanum*, I, 40.

72 Ray, *Methods plantarum nova* (London, 1682), 'Praefatio ad lectorem', p. 4.

73 Larson, *Reason and Experience* (Berkeley, Ca., 1971), p. 66.

74 Sloane, 'John Locke', p. 32, discussing *Methodus plantarum nova*, 'Praefatio', pp. 4–5.

75 My discussion here is fully dependent on Sloane, 'John Locke', p. 43. As Sloane points out, there is no direct evidence for Ray's adoption of Locke's philosophy, but the similarity of their arguments makes it quite likely. Sloane suggests the channels through which Ray may have become acquainted with Locke: one is William Charleton, a friend of Locke's who transmitted the 1687 abstract of his *Essay* from Holland to England. Ray had known Charleton from 1690 onwards. Charleton was a close friend of Ray's friend, Tancred Robinson. Second, the preparation for writing *The Wisdom of God* (1691) would lead Ray to familiarity with theological debates involving Locke's denial of innate ideas. Third, Ray was familiar with Locke's defender, William Sherlock, whose *Vindication of the Doctrine of the Holy and Ever Blessed Trinity* (London, 1690) adopts the Lockean position that the essences of things cannot be known. We might also point to Ray's and Locke's respective memberships in the Royal Society as well as their friendship with Wilkins. (See Sloane, 'John Locke', p. 39.)

76 J. P. Tournefort, *Élémens de botanique, ou méthode pour connoître les plantes* (Lyon, 1797), I, 70.

77 Ray, *Stirpium Europaearum extra Britannias nascentium sylloge* (London, 1694) in Sloane, 'John Locke', p. 35.

78 Ray to Robinson, 29 April 1696, in Ray, *Further Correspondence*, ed. Gunther, p. 301.

79 Ray, *De variis plantarum methodis dissertatio brevis*, in *Synopsis methodica stirpium Britannicarum*, 2nd edn (London, 1696), p. 1: 'Res singulares, quae solae, ut dixi, existunt, & sensus nostros feriunt, contemplati, earum multis maximam inter se similitudinem intervenire in omnibus ferè partibus & accidentibus, animadvertissent, iísque unum idémque nomen imposuissent; adeóq; res omnes quarum notitiam habuere, in quasdam quasi turmas divisissent, cujusmodi turmas Philosophi *Species infimas* vocant. Hujusmodi rerum in species saltem infimas divisio, & Nominum Generalium impositio, ad Linguae communis institutionem & usum omnino necessaria videtur.' In this and the following quotations, I have used Sloane's translations.

80 *Ibid.*, 'Praefatio', [A5v]: '...Quòd quamvis Universalia in rerum natura non existere, sed Intellectûs humani mera figmenta esse affirmaverim, fundamentum tamen in re habere non negem, nimirum convenientiam sue similitudinem in partibus & proprietatibus praecipuis.'

81 *Ibid.*, 'Praefatio', [A6r]: '...species hasce à se invicem essentialiter distingui, & intransmutabiles esse, earúmque formas sive essentias, vel principia quaedam sui

Generis esse, hoc est particulas quasdam minimas materiae, ab omnibus aliis distinctas & naturaliter indivisibiles, vel logos quosdam spermatikos idoneo sui generis veheculo inclusos.'

82 *Ibid.*, p. 5: 'Cùm enim omnis cognitio nostra originem ducat à sensu, nec quicquam de rebus quae extra nos sunt sciamus, quàm quód facultatem habeant hoc vel illo modo sensus nostros afficiendi, & mediantibus hisce impressionibus, talia vel talia phantasmata in Intellectu excitandi; si essentiae rerum sint formae immateriales, eas nullo modo sensibus nostris occurrere apud omnes in confesso est: Si verò nihil aliud sint quàm certa quaedam proportio & mixtura Principiorum sue minimorum naturalium; cùm minima illa nullo sensu nostro, quantumvis armato aut adjuto seorsim percipi possint... Cùm ergo essentiae rerum nobis incognitae sint, notae Generum essentiales à nobis certò sciri non possunt.'

83 *Ibid.*

84 *Ibid.*, p. 13: 'Naturâ convenire & Genere convenire idem est... Convenientiae in eadem Natura seu Genere nulla certior nota seu indicium esse potest quàm plura Attributa communia.'

85 See Stone, 'Literacy and education in England 1640–1900', *Past and Present*, 42 (1969) and 'The educational revolution in England 1560–1640', *Past and Present*, 28 (1964).

86 Margaret 'Espinasse, 'The decline and fall of Restoration science', *Past and Present*, 14 (1958), pp. 71–89.

87 *Ibid.*, p. 73.

88 Wilkins, *Mathematicall Magick* (London, 1648), vi.

89 C. R. Weld, *A History of the Royal Society* (London, 1898), I, 146, from Hooke's papers in British Museum Additional MSS 4441, dated 1663.

90 'Espinasse, 'The decline and fall', pp. 85ff. See also Margaret Jacob, *The Newtonians and the English Revolution 1689–1720* (Ithaca, N.Y., 1976), pp. 55ff.

91 Jacob, *Newtonians.*

92 'Espinasse, 'The decline and fall', p. 87.

93 John Locke, *Some Thoughts Concerning Education in Educational Writings*, ed. Adamson, p. 161.

94 Thomas Birch, *The History of the Royal Society* (London, 1756), I, 449-50.

95 *Ibid.*

96 Evelyn to Sir Peter Wyche (Chairman of the Royal Society Committee to Improve English), 20 June 1665, in Evelyn, *The Diary and Correspondence*, III, 309–12.

97 See the Preface to *Annus mirabilis* (1667) and the Dedication to *The Rival Ladies* (1664).

98 Thomas Sprat, *The History of the Royal Society* (London, 1667), p. 113.

99 Johnathan Swift, *A Proposal for Correcting, Improving and Ascertaining the English Tongue* (London, 1712), facsimile reprint Scolar Press (Menston, England, 1969). See also Swift's letter to the *Tatler*, no. 230, 28 September 1710, in *The Prose Works of Jonathan Swift*, ed. Herbert Davis (Oxford, 1968), II, 173–7.

100 Locke, *Essay*, 3.9.15.

101 See Hugh Kenner, 'Pope's reasonable rhymes', *English Literary History*, 41 (1974), pp. 74–88.

102 Joseph Addison, *Spectator*, July 1711, in 'Espinasse, 'The decline and fall', p. 84.

Bibliography

Aaron, Richard I. *John Locke*, 3rd edn. Oxford, 1971

Aarsleff, Hans. 'Alsted', *Dictionary of Scientific Biography*, I, 359–63

'Leibniz on Locke on Language', *American Philosophical Quarterly*, I (1964), pp. 165–88

'Wilkins', *Dictionary of Scientific Biography*, XIV, 361–81

Albertus Magnus. *The Book of Secrets of Albertus Magnus of the Virtues of Herbs, Stones and Certain Beasts, also a Book of the Marvels of the World*. Ed. Michael R. Best and Frank H. Brightman. Oxford, 1973

Aldrovandi, Ulisse. *Aldrovandi on Chickens: The Ornithology of Ulisse Aldrovandi (1600), Volume II, Book xiv*. Trans. L. R. Lind. Norman, Okla., 1963

Alston, R. C. A. *The Bibliography of the English Language from the Invention of Printing to the year 1800*. 11 vols. Leeds, 1965–73

Anderson, Frank J. *An Illustrated History of the Herbals*. New York, 1977

Anderson, Fulton H. *The Philosophy of Francis Bacon*. Chicago, 1948

Andrade, C. N. de C. 'The Real Character of Bishop Wilkins', *Annals of Science*, I (1936), pp. 4–12

Arber, Agnes. *Herbals: Their Origin and Evolution*. Cambridge, 1938

Aristotle, *On Interpretation*. Trans. Harold P. Cook. Cambridge, Ma., 1962

Posterior Analytics. Trans. Hugh Tredennick. Cambridge, Ma., 1961

The Works of Aristotle. Ed. W. D. Ross. 12 vols. London, 1908–52

Ashmole, Elias, trans. *Fasciculus chemicus* by Arthur Dee. London, 1650

Aubrey, John. *Brief Lives*. Ed. Oliver Dick. London, 1958

Brief Lives. Ed. Andrew Clark. 2 vols. Oxford, 1898

Aubrey on Education. Ed. J. E. Stephens. London, 1972

Bacon, Francis. *The Work of Sir Francis Bacon*. Ed. and trans. James Spedding and Robert Ellis. 14 vols. 1857–74; rpt. New York, 1968

Baker, Thomas. *Reflections Upon Learning*. London, 1700

Balme, D. M. 'Genos and eidos in Aristotle's biology', *Classical Quarterly*, 12 n.s. (1962), pp. 81–98

Banckes. *An Herbal 1525*. Ed. and transcribed into modern English with an Introduction by Stanford V. Larkey and Thomas Pyles. New York, 1941

Barnett, Pamela. 'Theodore Haak and the early years of the Royal Society', *Annals of Science*, XIII (1957), pp. 205–18

Theodore Haak, F.R.S. The Hague, 1962

Bartholomaeus Anglicus. *On the Properties of Things: John Trevisa's Translation of*

Bartholomaeus Anglicus De Proprietatibus Rerum. Ed. M. C. Seymour, et al. 2 vols. Oxford, 1975–

Bauhin, Caspar. *Pinax theatri botanici*. 2nd edn. Basel, 1671

Becher, Johann Joachim. *Spirensis character pro notitia linguarum universali, inventum steganographicum*. Frankfurt, 1661

Beck, Cave. *The Universal Character, By Which all Nations in the World may understand one Another's Conceptions, Reading out of one Common Writing their Own Mother Tongues*. London, 1657

Berlin, Brent. 'Speculations on the growth of ethnobotanical nomenclature', *Language in Society*, 1 (1972), pp. 51–86

Berlin, B., Breedlove, Dennis E., Raven, Peter H. 'General principles of classification and nomenclature in folk biology', *American Anthropologist*, 75 (1973), pp. 214–42

Bermudo, Pedro. *Arithmeticus nomenclator mundi omnes nationes ad linguarum et sermonis imitatem invitans*. Rome, 1653

Biographie universelle. Ed. M. M. Michaud. 2nd edn. 45 vols. Paris, 1880

Birch, Thomas. *The History of the Royal Society*. London, 1756

The Life of Tillotson. London, 1752

Bloomfield, Leonard. *Language*. New York, 1965

Blundeville, Thomas. *The Art of Logicke*. London, 1599

Borges, Jorge Luis. *Other Inquisitions*. Trans. Ruth L. C. Simms. 1964; rpt. New York, 1968

Boyle, Robert. *The Works of the Honourable Robert Boyle*. Ed. Thomas Birch. 6 vols. London, 1772

Bradley, Henry. *On the relations between spoken and written language with special references to English*. A paper read at the International Historical Congress, April 1913. London, n.d.

Burnet, Gilbert. *A History of My Own Time*. 2 vols. London, 1833

Carré, Meyrick H. *Phases of Thought in England*. Oxford, 1949

Cassidy, John Robert. 'Aristotle on definitions', *Southern Journal of Philosophy*, V (1967), pp. 110–18

Cesalpino, Andrea. *De plantis libri XVI*. Florence, 1583

Questions Péripatéticiennes. Trans. Maurice Dorolle. Paris, 1929

Charleton, Walter. *The Immortality of the Human Soul*. London, 1657

Clark, Herbert E. and Eve V. *Psychology and Language*. New York, 1977

Cohen, Murray. *Sensible Words: Linguistic Practice in England 1640–1785*. Baltimore, Md, 1977

Cole, Percival. *A Neglected Educator: Johann Heinrich Alsted*. Sydney, 1910

Coles, William. *Adam in Eden or Nature's Paradise: The History of Plants, Fruits, Herbs and Flowers*. London, 1657

Collingwood, R. G. *The Idea of Nature*. 1945; rpt. Oxford, 1965

Comenius, John Amos. *Analytic Didactic*. Ed. and trans. Vladimir Jelinek. Chicago, 1953

The Great Didactic. Ed. and trans. M. W. Keatinge. London, 1896

Janua linguarum reserata. Leszno, 1631

Janua linguarum reserata: The Gate of Languages Unlocked. Trans. F. R. Horn. Corrected by J. Robotham. London, 1643

Janua linguarum reserata vestibulum. Leszno, 1633
Janua rerum reserata sive universalis sapientia. Ed. Klaus Schaller. Munich, 1968
Natural Philosophy Reformed by Divine Light or a Synopsis of Physics. London, 1651
Orbis sensualium pictus. The Visible World: or a picture and nomenclature of all the chief things that are in the world and of men's employments therein. Trans. Charles Hoole. (3rd edn, London, 1672). Introduction by James Bowen. Sydney, 1967
Panglottia in *De rerum emendatione consultatio catholica.* 2 vols. Prague, 1966. II, 147–204
A Pattern of Universal Knowledge...or a Diatyposis or Model of the Eminently Learned and Pious Promoter of Science in Generall. Trans. Jeremy Collier. London, 1662
Via lucis. Ed. and trans. E. T. Campagnac. Liverpool, 1938
Cornelius, Paul. *Language in Seventeenth- and Early Eighteenth-Century Imaginary Voyages.* Geneva, 1965
Couturat, Louis. *La Logique de Leibniz.* 1901; rpt. Hildesheim, 1961
Opuscules et fragments inédits de Leibniz. Paris, 1903
Couturat, L. and Leau, L. *Histoire de la langue universelle.* Paris, 1903
Cram, David. 'George Dalgarno on "Ars Signorum" and Wilkins' "Essay"', in *Progress in Linguistic Historiography.* Ed. Konrad Koerner. Amsterdam Studies in the Theory and History of Linguistic Science: series 3, Studies in the History of Linguistics. Vol. 20. Amsterdam, 1980. pp. 113–21
Cranston, Maurice. *John Locke, A Biography.* New York, 1957
Cressy, David. *Literacy and the Social Order: Reading and Writing in Tudor and Stuart England.* Cambridge, 1980
Crombie, A. C. *Augustine to Galileo.* 2 vols. Harmondsworth, 1969
Dalgarno, George. *Ars signorum.* London, 1661
'A New Discovery of the Universal Character, Containing Also a More Readie and Approved Way of Short-Hand Writting, then any heretofore practised in this Nation'. Oxford, 1657. Sloane MS 4377, f. 144
'News to the Whole World, of the Discovery of an Universal Character, and a New Rational Language'. Sloane MS 4377, f. 143
'Omnibus omnio hominibus orbis terrarum praecipue Dominis'. Sloane MS 4377, ff. 139–42
'The possibilitie and great usefulnesse of an Universall character'. Sloane MS 4377, f. 157
'The Tables of the Universal Character'. Oxford, 1657. Sloane MS 4377, ff. 145–6
The Works of George Dalgarno of Aberdeen. A Maitland Club Publication. Edinburgh, 1834
Davies, H. Neville. 'Godwin's "Lunatique Language"', *Journal of the Warburg and Cortauld Institute,* 30 (1967), pp. 296–316
Debus, A. *The English Paracelsians.* New York, 1965
Science and Education in the Seventeenth Century: the Webster–Ward Debate. London, 1970
ed., *Medicine in England.* Berkeley, Ca., 1974
De Mott, Benjamin. 'Comenius and the Real Character in England', *Publications of the Modern Language Association,* LXX (1955), pp. 1068–81
'Science versus mnemonics', *Isis,* 48 (1957), pp. 3–12

'The sources and development of John Wilkins' philosophical language', *Journal of English and Germanic Philology*, LVII (1958), pp. 1–12

Descartes, René. *Oeuvres Complètes de Descartes*. Ed. Charles Adams and Paul Tannery. 12 vols. Paris, 1897–1910

Descartes' Philosophical Letters, Ed. Anthony Kenny. Oxford, 1970

Philosophical Works of Descartes. Trans. E. Haldane and G. Ross. 2 vols. Cambridge, 1931

Dewhurst, Kenneth. *John Locke (1632–1704), Physician and Philosopher: A Medical Biography*. London, 1963

Dictionary of National Biography. Ed. Sir Leslie Stephen and Sir Sidney Lee. 22 vols. Oxford, 1921–2

Dictionary of Scientific Biography. Ed. C. C. Gillispie. 15 vols. New York, 1970–80

Digby, Kenelm. *Two treatises. In the one of which, the nature of bodies; in the other the nature of man's soul; is looked into*... Paris, 1643

Dioscorides. *De materia medica: libri quinque*. Ed. Max Wellmann. Berlin, 1958

The Greek Herbal of Dioscorides. Ed. R. T. Gunther. 1933; rpt. Oxford, 1934

Du Moulin, Pierre. *The Elements of Logick written first in Latine, then in French... and now translated into English*. Trans. Joshua Ahier. Oxford, 1647

Eisenstein, Elizabeth. *The Printing Press as an Agent of Change: Communications and Cultural Transformations in Early Modern Europe*. 2 vols. Cambridge, 1979

Elliott, Ralph W. V. 'Isaac Newton as phonetician', *Modern Language Review*, 49 (1954), pp. 5–12

'Isaac Newton's "Of an Universall Language"', *Modern Language Review*, 52 (1957), pp. 1–18

Encyclopedia of Philosophy. Ed. Paul Edwards. 8 vols. New York, 1967

'Espinasse, Margaret. 'The decline and fall of Restoration science', *Past and Present*, 14 (1958), pp. 71–89

Robert Hooke. London, 1956

Evelyn, John. *The Diary of John Evelyn, esq., F.R.S.* Ed. William Bray. 4 vols. London, 1906

FitzMaurice, Edmond. *The Life of Sir William Petty*. London, 1895

Foucault, Michel. *The Order of Things: An Archaeology of the Human Sciences*. London, 1970

Fox-Bourne, Henry R. *The Life of John Locke*. 2 vols. London, 1876

Foxcroft, H. C. *A Supplement to Burnet's History of My Own Time*. Oxford, 1902

Frank, Robert G. Jr. 'Physician as virtuoso in seventeenth century England', in *English Scientific Virtuosi in the 16th and 17th Centuries*. Papers read at a Clark Library Seminar, 5 February 1977. Los Angeles, Ca., 1979, pp. 59–123

Gerard, John. *The Herbal or General Historie of Plants*. London, 1597

Givner, David. 'Scientific preconceptions in Locke's philosophy of nature', *Journal of the History of Ideas*, 23 (1962), pp. 340–54

Godwin, Francis. *The Man in the Moone: or a Discourse of a Voyage Thither by Domingo Gonsales, the Speedy Messenger* (1638). Ed Grant McColley. *Smith College Studies in Modern Languages*, 19 (1937), pp. 1–48

Nuncius inanimatus (1629). Ed. Grant McColley. *Smith College Studies in Modern Languages*, 19 (1937), pp. 50–67

Golding, Arthur, trans. *A Woorke concerning the Trewnesse of the Christian Religion*. London, 1587

Goody, Jack. *The Domestication of the Savage Mind*. Cambridge, 1977

Greene, E. L. 'Botanical literature old and new', *Pittonia*, I (1888), pp. 176–94

Landmarks of Botanical History. Smithsonian Miscellaneous Collections. Vol. 54. Washington, D.C., 1909

Grene, Marjorie. *A Portrait of Aristotle*. Chicago, 1963

Grene, M. and Mendelsohn, E. *Topics in the Philosophy of Biology*. Boston Studies in the Philosophy of Science. Vol. 27. Boston, 1976

Gunther, R. W. T. *Early Science in Oxford*. 14 vols. Oxford, 1928

Hall, A. Rupert. 'Science, technology and utopia', in *Science and Society 1600–1900*. Ed. Peter Mathias. Cambridge, 1972

The Scientific Revolution. 1954; rpt. London, 1967

Harrison, J. and Laslett, P. *The Library of John Locke*. Oxford, 1971

Hart, John. *A Methode or comfortable beginning for all unlearned, whereby they may be taught to read English, in a very short time, with pleasure*. London, 1570

Hartlib, Samuel. *Ephemerides*. Sheffield University Library

A Reformation of the Schools. London, 1642

Samuel Hartlib and the Advancement of Learning. Ed. Charles Webster. Cambridge, 1970

Havelock, Eric A. *Origins of Western Literacy*. Ontario Institute for Studies in Education Monograph Series/14. Toronto, 1976

Preface to Plato. Cambridge, Ma., 1963

Prologue to Greek Literacy. Norman, Okla., 1973

Havránek, Bohuslav. 'The functional differentiation of the standard language' (1932) in *A Prague School Reader on Esthetics, Literary Structure and Style*. Ed. and trans. Paul L. Garvin. Washington, D.C., 1964

Heller, John L. 'The early history of binomial nomenclature', *Huntia*, I (1964), pp. 33–68

Henrey, Blanche. *British Botanical and Horticultural Literature before 1800*. 3 vols. Oxford, 1976

Hesse, Mary B. 'Francis Bacon's Philosophy of Science', in *A Critical History of Western Philosophy*. Ed. D. J. O'Connor. New York, 1964. Reprinted in *Essential Articles for the Study of Francis Bacon*. Ed. Brian Vickers. Hamden, Conn., 1968, pp. 114–39

'Hooke's philosophical algebra', *Isis*, 57 (1966), pp. 67–83

Hill, Abraham. *Familiar Letters*. London, 1767

Hill, Christopher. *The World Turned Upside Down: Radical Ideas During the English Revolution*. Harmondsworth, 1975

Hirsch, E. D. *The Philosophy of Composition*. Chicago, 1977

Hobbes, Thomas. *The English Works of Thomas Hobbes*. Ed. Sir William Molesworth. 11 vols. London, 1839–45

Hooke, Robert. *A Description of Helioscopes and some other instruments*. London, 1676

The Diary of Robert Hooke. Ed. Henry Robinson and Walter Adams. London, 1935

'Mathematical language'. Royal Society Classified Papers. Vol. 20, no. 72

The Posthumous Works of Robert Hooke (London, 1705). Ed. Richard S. Westfall. New York, 1969

Huygens, Christian. *Oeuvres Complètes*. Société Hollandaise des Sciences. 22 vols. The Hague, 1888–1950

Ivins, William. *Prints and Visual Communication*. Cambridge, Ma., 1953

Jacob, J. R. *Robert Boyle and the English Revolution*. New York, 1977

Jacob, Margaret C. *The Newtonians and the English Revolution 1689–1720*. Ithaca, N.Y., 1976

Jardine, Lisa. *Francis Bacon: Discovery and the Art of Discourse*. Cambridge, 1967

Joachim, Harold. *Descartes's Rules for the Direction of the Mind, by Harold H. Joachim; reconstructed from notes taken by his pupils*. Ed. Errol Harris. London, 1957

Jones, R. F. *Ancients and Moderns*. 1961; rpt. Berkeley, Ca., 1965

The Triumph of the English Language. 1953; Palo Alto, Ca., 1966

Kangro, Hans. *Joachim Jungius' Experimente und Gedanken zur Begründung der Chemie als Wissenschaft*. Wiesbaden, 1968

Kargon, Robert. *Atomism in England from Hariot to Newton*. Oxford, 1966

Kearney, H. F. 'Puritanism, capitalism and the scientific revolution', in *The Intellectual Revolution of the Seventeenth Century*. Ed. C. Webster. London, 1974

Kenner, Hugh. 'Pope's reasonable rhymes', *English Literary History*, 41 (1974), pp. 75–80

Kinder, Philip. 'Philip Kynder's Book', Bodleian Library Ashmole MS 788

Kinner, Cyprian. *A Continuation of J. A. Comenius School Endeavours*. London, 1648

Kircher, Athanasius. *Polygraphia nova et universalis ex combinatoria arte detecta*. Rome, 1663

Kneale, William and Martha. *The Development of Logic*. Oxford, 1962

Knowlson, James. *Universal Language Schemes in England and France 1600–1800*. Toronto, 1975

'The ideal languages of Veiras, Foigny, and Tyssot de Patot', *Journal of the History of Ideas*, 24 (1963), pp. 269–78

'Jean Le Maire, the Almérie, and the "Musique Almérique"', *Acta Musicologica*, 40 (1968), pp. 86–9

'A note on Bishop Godwin's *Man in the Moone:* The East Indies Trade Route and a "Language of Musical Notes"', *Modern Philology*, 65 (1965), pp. 357–61

Kretzmann, Norman. 'The main thesis of Locke's semantic theory', *Philosophical Review*, 77 (1968), pp. 175–96

Land, Stephen K. *From Signs to Propositions*. London, 1974

Larson, James. *Reason and Experience*. Berkeley, Ca., 1971

Leeuwen, Henry G. Van. *The Problem of Certainty in English Thought 1630–1690*. The Hague, 1963

Leibniz, Gottfried Wilhelm. *Die Philosophischen Schriften von Gottfried Wilhelm Leibniz*. Ed. C. I. Gerhardt. 7 vols. Berlin, 1890; rpt. Hildesheim, 1961

Lenoble, Robert. *Mersenne ou la Naissance du Mécanisme*. Paris, 1943

Lever, Ralph. *The Reason, rightly termed Witcraft, teaching a perfect way to argue and dispute*. London, 1573

Lévi-Strauss, Claude. *The Savage Mind*. 1962; rpt. Chicago, 1968

Lewis, C. S. *The Discarded Image: An Introduction to Medieval and Renaissance Literature*. Cambridge, 1964

Lewis, Mark. *An Apologie for Grammar*. London, 1671

An Essay to Facilitate the Education of Youth, by Bringing Down the Rudiments of Grammar to the Sense of Seeing. London, 1670(?)

Linnaeus, Carolus. *Genera plantarum*. 5th edn (1754). Ed. W. T. Stearn. Weinheim, 1960

Species plantarum. 2 vols. Holmiae, 1753

Systema naturae. 12th edn. Holmiae, 1776

Locke, John. *The Correspondence of John Locke*. Ed. E. S. De Beer. 4 vols. Oxford, 1976–

An Early Draft of Locke's Essay together with Exerpts from his Journals. Ed. R. I. Aaron and J. Gibb. Oxford, 1936

The Educational Writings of John Locke. Ed. John W. Adamson. 2nd edn. Cambridge, 1922

The Educational Writings of John Locke. Ed. James L. Axtell. Cambridge, 1968

An Essay Concerning Human Understanding. Ed. A. C. Fraser. 2 vols. Oxford, 1894

Essays on the Laws of Nature. Ed. W. Von Leyden. Oxford, 1968

Locke's Travels in France. Ed. John Lough. Cambridge, 1965

Lodowyck, Francis. *A Common Writing whereby two although not understanding one the other's Language, yet by the helpe thereof may communicate their minds one to another*. London, 1647

'Concerning Mr. Dalgarno's universall carracter and language' (1657). Sloane MS 932, ff. 13v–15v

'An Essay Towards an Universal Alphabet and Universal Primer', *Philosophical Transactions of the Royal Society*, 16, no. 182 (1686), pp. 126–33

The Groundwork and Foundation Laid (or so Intended) For the Framing of a New Perfect Language: And an Universal or Common Writing. London, 1652

'A Second Essay Concerning a Universal Primer', *Philosophical Transactions of the Royal Society*, 16, no. 182, (1686), pp. 134–7

'Of an universall reall caracter', (1657). Sloane MS 897, ff. 32r–39v

Loemaker, Leroy E. 'Leibniz and the Herborn Encyclopedists', *Journal of the History of Ideas*, 22 (1961), pp. 323–38

Louis, A. 'Les Dénominations binomiales des plantes dans l'antiquité grecque', *Scientiarum Historia*, 13 (1971), pp. 1–26

Lovejoy, Arthur O. *The Great Chain of Being*. 1936; rpt. Cambridge, Ma., 1973

Mandelbaum, Maurice. *Philosophy, Science and Sense Perception*. Baltimore, Md., 1964

Manuel, Frank. *A Portrait of Isaac Newton*. Cambridge, Ma., 1968

Maritain, Jacques. *An Introduction to Philosophy*. Trans. E. I. Watkin. London, 1930

Menninger, Karl. *Number Words and Number Symbols*. Trans. Paul Broneer. Cambridge, Ma., 1969

Mersenne, Marin. *Correspondance du Marin Mersenne*. Ed. Cornélis de Waard. 12 vols. Paris, 1932–

Harmonie universelle. Paris, 1636

Merton, R. K. *Science, Technology and Society in 17th Century England*. 1938; rpt. New York, 1970

Michael, Ian. *English Grammatical Categories and the Tradition to 1800*. Cambridge, 1970

Miškovskà, V. T. 'Comenius (Komenský) on lexical symbolism in an artificial language', *Philosophy*, 37 (1962), pp. 238–44

'Empirisme ou invention à la base lexicale de la Panglottia?' *Acta Comeniana*, 20 (1961), pp. 189–97

'La Panglottie de J. A. Komenský', *Philologica Pragensia*, II (1959), pp. 97–105

'Langue philosophique et religion au XVIIe siècle', *Communio Viatorum*, 2 (1959), pp. 335–42

'La Triertium et la Panglottie', *Acta Comeniana*, 3 (1973), pp. 383–90

Moravcsik, J. M. E. 'Aristotle's theory of categories', in *Aristotle: A Collection of Critical Essays*. Ed. J. M. E. Moravcsik. Garden City, N.Y., 1967

Newton, Isaac. *Correspondence*. Ed. H. W. Turnbull. 3 vols. Cambridge, 1959

Nicolson, M. H. *Conway Letters 1642–1684*. Oxford, 1930.

Oldenburg, Henry. *The Correspondence of Henry Oldenburg*. Ed. A. Rupert Hall and Marie Boas Hall. 9 vols. Madison, Wis., 1968–

O'Neil, Brian E. *Epistemological Direct Realism in Descartes' Philosophy*. Albuquerque, N.M., 1974

Ong, Walter. 'From allegory to diagram in the Renaissance mind', *Journal of Aesthetics and Art Criticism*, XVII (1959), pp. 423–40

Interfaces of the Word: Studies in the Evolution of Consciousness and Culture. Ithaca, N.Y., 1977

Ramus, Method, and the Decay of Dialogue. Cambridge, Ma., 1958

Rhetoric, Romance, and Technology: Studies in the Interaction of Expression and Culture. Ithaca, N.Y., 1971

'System, space and intellect in Renaissance symbolism', *Bibliothèque d'Humanisme et Renaissance*, 8 (1956), pp. 222–39

Owen, John. *Of the Divine Original*. London, 1659

Owens, J. 'Aristotle on categories', *Review of Metaphysics*, 14 (1960–1), pp. 73–90

Padley, G. A. *Grammatical Theory in Western Europe 1500–1700: The Latin Tradition*. Cambridge, 1976

Parkinson, John. *Theatrum Botanicum: The Theater [sic] of Plants or An Herball of a Large Extent*. London, 1640

Paschall, Andrew. *Propositions for Carrying on a Philosophical Correspondence, already begun in the County of Somerset upon incouragement given from the Royal Society*. London, 1670

Pears, David. 'Grasping the essentials'. *Times Literary Supplement*. 13 February 1976, p. 174

Pepys, Samuel. *The Diary of Samuel Pepys*. Ed. H. B. Wheatley. 2 vols. London, 1893; rpt. New York, 1946

Letters and the Second Diary of Samuel Pepys. Ed. R. G. Howarth. London, 1932

Petty, William. *The Advice of W.P. to Mr. Samuel Hartlib, For the Advancement of Some Particular Parts of Learning*. London, 1648

Petty Papers. Ed. Marquis of Landsdowne. 2 vols. London, 1927

Plot, Robert. *The Natural History of Oxfordshire, being an Essay toward the Natural History of England*. Oxford, 1676

Pons, E. 'Les Langues imaginaires dans le voyage utopique: les deux grammairiens: Vairasse et Foigny', *Revue de Littérature Comparée*, 12 (1932), pp. 500–32

Popper, Karl R. *Conjectures and Refutations: the Growth of Scientific Knowledge*. 1963; New York, 1968

The Open Society. 4th edn rev. 2 vols. Princeton, N.J., 1963

Porphyry. *The Organon or Logical Treatises of Aristotle with the Introduction of Porphyry.* Ed. and trans. Octavius F. Owen. 2 vols. London, 1853

Porphyrii isagoge et in Aristotelis categorias commentarium. Ed. A. Busse. Berlin 1887, in *Commentaria in Aristotelem Graeca,* Academiae Litterarum Regiae Borusicae (Berlin, 1887), IV (i)

Purver, Margery. *The Royal Society: Concept and Creation.* London, 1967

Rattansi, P. M. 'The social interpretation of science', in *Science and Society 1600–1900.* Ed. Peter Mathias. Cambridge, 1972

Raven, Charles. *John Ray Naturalist, His Life and Works.* Cambridge, 1942

Ray, John. *Catalogus plantarum Angliae.* London, 1670 (2nd edn 1677)

Catalogus plantarum circa Cantabrigiam nascentium. Cambridge, 1660

A Collection of English Proverbs. Cambridge, 1670

A Collection of Words not Generally Used. London, 1674

The Correspondence of John Ray. Ed. Edwin Lankester. London, 1848

A Discourse on the Specific Differences of Plants (1674) in Ray, *Further Correspondence.* Ed. R. W. T. Gunther, pp. 77–83

Further Correspondence of John Ray. Ed. R. W. T. Gunther. London, 1928

Historia plantarum. London, 1686

Memorials of John Ray consisting of his life by Dr. Derham. London, 1846

Methodus plantarum emendata et aucta. London, 1703

Methodus plantarum nova. London, 1682

Nomenclator classicus sive dictionariolum trilingue. London, 1703

Philosophical Letters Between the late learned Mr. Ray and Several of His ingenious Correspondents. Ed. William Derham. London, 1718

Synopsis methodica stirpium Britannicarum. London, 1690 (2nd edn 1696)

The Wisdom of God Manifested in the Works of Creation. London, 1691

Reif, Patricia. 'The textbook tradition in natural philosophy 1600–1650', *Journal of the History of Ideas,* 30 (1969), pp. 17–32

Renaudaut, Théophraste. *A General Collection of the discourses of the Virtuosi of France.* A translation of the first one hundred conferences of the 'Recueil général des questions traitées ès conférences du Bureau d'Adresse'. 5 vols. Paris, 1638–55. Trans. G. Havers. London, 1664

Rigaud, Stephen, ed. *Correspondence of Scientific Men of the Seventeenth Century.* 2 vols. Oxford, 1841

Romanell, Patrick. 'Locke and Sydenham: A Fragment on Smallpox (1670)', *Bulletin of the History of Medicine,* XXXII (1958), p. 316

Ross, W. D. *Aristotle.* 1923; rpt. London, 1960

Rossi, Paolo. *Clavis unversalis.* Milan, 1960

Francis Bacon. Trans. Sacha Rabinovitch. London, 1968

Royal Society of London. *Philosophical Transactions 1665/6–1750.* 46 vols. London.

Ryle, Gilbert. 'Categories', in *Logic and Language.* 2nd series. Ed. Anthony Flew. Oxford, 1953

Sachs, Julian von. *History of Botany 1530–1860.* Trans. Henry Garnseey. Oxford, 1906

Sadler, Edward. *J. A. Comenius and the Concept of Universal Education.* London, 1966

Salmon, Vivian. 'The Evolution of Dalgarno's "Ars Signorum"', in *Studies in Language and Literature in Honour of Margaret Schlauch.* Ed. M. H. Brahmer, et al. Warsaw, 1966, pp. 353–73

'John Wilkins' *Essay* (1668): critics and continuators', *Historiographia Linguistica*, I (1974), pp. 147–63

'Joseph Webbe: some seventeenth century views on language teaching and the nature of meaning', *Bibliothèque d'Humanisme et Renaissance*, 23 (1961), pp. 324–40

'Language planning in 17th Century England: its contexts and aims', in *In Memory of J. R. Firth*. Ed. Charles F. Bazell. London, 1966, pp. 370–97

'Problems of language teaching: a discussion among Hartlib's friends', *Modern Language Review*, LIX (1964), pp. 13–24

Rev. art. of *Cartesian Linguistics* by Noam Chomsky in *Journal of Linguistics*, 5 (1969), pp. 165–87

The Works of Francis Lodwick: A study of his writings in the intellectual context of the 17th century. London, 1972

Scheu, Sister Marina. *Categories of Being in Aristotle and St. Thomas*. Washington, D.C., 1944

Schofield, Roger. 'Dimensions of illiteracy 1750–1850', *Explorations in Economic History*, 10 (1973), pp. 437–54.

Schott, Gaspar. *Technica curiosa, sive mirabilia artis. libris XII, comprehensa, &c.* Nuremburg, 1664

Seeber, E. 'Ideal languages in the French and imaginary voyages', *Publications of the Modern Language Association*, LX (1945), pp. 586–97

Shapiro, Barbara. 'History and natural history in sixteenth- and seventeenth-century England: an essay on the relationship between humanism and science', in *English Scientific Virtuosi in the 16th and 17th Centuries*. Papers read at a Clark Library Seminar, 5 February 1977. Los Angeles, 1979, pp. 3–55

John Wilkins: An Intellectual Biography 1614–1672. Berkeley, Ca., 1969

Shuckburgh, E. S., ed. *Two Biographies of William Bedell*. Cambridge, 1902

Simpson, William. *Hydrologia chymica*. London, 1669

Sloane, Philip. 'John Locke, John Ray, and the problem of natural system', *Journal of the History of Biology*, 5 (1972), pp. 1–53

Somerset, Edward (Earl of Worcester). *A Century of Names and Scantlings*. Ed. Charles F. Partington. London, 1825

Sprague, Thomas A. *The Evolution of Botanical Taxonomy from Theophrastus to Linnaeus*. Linnean Society of London. London, 1950

Sprat, Thomas. *The History of the Royal Society* (London, 1667). Ed. Jackson I. Cope and Harold Whitmore Jones. St Louis, Mo., 1958

Stafleau, Frans. *Linnaeus and the Linnaeans*. Utrecht, 1971

Stearn, W. T. 'The background of Linnaeus' contribution to the nomenclature and methods of systematic biology', *Systematic Zoology*, 8 (1959), pp. 4–22

Stewart, Dugald. *The Collected Works of Dugald Steward*. Ed. Sir William Hamilton. 2 vols. Edinburgh, 1854

Stimson, Dorothy. *Science and Amateurs*. New York, 1946

Stone, Lawrence. 'The educational revolution in England 1560–1640', *Past and Present*, 28 (1964), pp. 41–80

'Literacy and education in England 1640–1900', *Past and Present*, 42 (1969), pp. 69–139

Sturm, Johann Christopher. *Collegium experimentale sive curiosum*. Nuremburg, 1676

Swift, Jonathan. *A Proposal for Correcting, Improving and Ascertaining the English Tongue.* London, 1712. Fac. rpt. Scolar Press. Menston, England, 1979

The Prose Works of Jonathan Swift. Ed. Herbert Davis. 14 vols. Oxford, 1968

Syfert, R. H. 'The origins of the Royal Society', *Notes and Records of the Royal Society,* V (1948), pp. 75–137

Theophrastus. *Enquiry into Plants.* Trans. Sir Arthur Hort. 2 vols. London, 1916

Thomas, Keith. *Religion and the Decline of Magic.* Harmondsworth, 1973

Tournefort, J. P. *Élémens de botanique, ou méthode pour connoître les plantes.* 6 vols. Lyon, 1797

Trevor-Roper, Hugh. *Religion, Reformation and Social Change.* London, 1972

Turnbull, G. H. *Hartlib, Dury and Comenius.* Liverpool, 1947

'Samuel Hartlib's influence on the early history of the Royal Society', *Notes and Records of the Royal Society,* X (1953), pp. 101–30

Turner, A. J. 'Andrew Paschall's table of plants for the universal language (1678)', *Bodleian Library Record,* LX (1978), pp. 346–50

Ullmann, Stephen. 'Semantic Universals', in *Universals of Language.* 2nd edn. Ed. J. H. Greenberg. Cambridge, Ma., 1963

Urquhart, Thomas. *The Works of (Sir) Thomas Urquhart.* A Maitland Club Edition. Edinburgh, 1834

Váczy, C. 'Les Origines et les principes du developpement de la nomenclature binaire en botanique', *Taxon,* 20 (1971), pp. 573–90

Vaughan, Robert. *The Protectorate of Oliver Cromwell.* 2 vols. London, 1838

Vines, S. H. 'Robert Morrison and John Ray', in *Makers of British Botany.* Ed. F. W. Oliver. Cambridge, 1913

Wallis, John. *A Defence of the Royal Society and Philosophical Transactions...in Answer to the Cavils of Dr. William Holder.* London, 1678

'Letter of Dr. John Wallis to Robert Boyle, Esq; Concerning the said Doctor's Essay Teaching a person Dumb and Deaf to Speak and to Understand Language', *Philosophical Transactions of the Royal Society,* 5, no. 61 (1670), p. 1092.

Ward, Seth. *Vindiciae academiarum.* Oxford, 1654

Webb, John. *An historical Essay endeavoring a Probability that the Language of the Empire of China is the Primitive Language.* London, 1669

Webster, Charles. 'English medical reformers of the Puritan Revolution', *Ambix,* XIV (1967), pp. 16–41

The Great Instauration: Science, Medicine and Reform 1626–1660. New York, 1976

'Marcaria: Samuel Hartlib and the Great Reformation', *Acta Comeniana,* 2 (1971), pp. 147–63.

ed., *Samuel Hartlib and the Advancement of Learning.* Cambridge, 1970

Webster, John. *Academiarum Examen, or the Examination of the Academies.* London, 1654

Weld, C. R. *A History of the Royal Society.* 2 vols. London, 1898

Whitaker, Virgil. 'Francis Bacon's intellectual milieu', in *Essential Articles for the Study of Francis Bacon.* Ed. Brian Vickers. Hamden, Conn., 1968, pp. 28–50

Whitehead, Alfred North. *Science and the Modern World.* 1925; rpt. New York, 1967

Wilkins, John. *An Essay Towards a Real Character and a Philosophical Language.* London, 1668

The Mathematical and Philosophical Works of the Right Reverend John Wilkins. 2 vols. London, 1802

Mathematicall Magick. Or, The Wonders that may be performed by Mechanicall Geometry. London, 1648

Mercury; Or, the Secret and Swift Messenger: Shewing How a Man May with Privacy and Speed Communicate His Thoughts to a Friend at Any Distance. London, 1641

Of the Principles and Duties of Natural Religion. London, 1675

Wilson, John Cook. 'Categories in Aristotle and in Kant', in *Aristotle: A Collection of Critical Essays.* Ed. J. M. E. Moravcsik. Garden City, N.Y., 1967

Wilson, Thomas. *Rule of Reason.* London, 1551

Wood, Anthony. *Athenae oxonienses.* 2nd edn. 2 vols. London, 1721

Worthington, John. *The Diary and Correspondence of Dr. John Worthington.* Ed. J. Crossley. 2 vols. London, 1847

Wotton, William. *Reflections on Ancient and Modern Learning.* London, 1694

Yates, Frances. *The Art of Memory.* Harmondsworth, 1969

The Rosicrucian Enlightenment. London, 1972

Yolton, John W. *John Locke and Education.* New York, 1971

Locke and the Compass of Human Understanding. Cambridge, 1970

Young, Robert. *A Bohemian Philosopher at Oxford in the 17th Century.* London, 1925

Comenius in England. London, 1932

Index

For EU product safety concerns, contact us at Calle de José Abascal, 56–1°, 28003 Madrid, Spain or eugpsr@cambridge.org.

www.ingramcontent.com/pod-product-compliance
Lightning Source LLC
LaVergne TN
LVHW090314210926
848480LV00001B/247